Lecture Notes in Mathematics 1616

Springer
Berlin
Heidelberg
New York
Barcelona
Budapest
Hong Kong
London
Milan
Paris
Santa Clara
Singapore
Tokyo

I. Moerdijk

Classifying Spaces and Classifying Topoi

Springer

Author

Izak Moerdijk
Mathematisch Institut
Universiteit Utrecht
P.O. Box 80.010
3508 TA Utrecht, The Netherlands

Cataloging-in-Publication Data applied for

Die Deutsche Bibliothek - CIP-Einheitsaufnahme

Moerdijk, Izak:
Classifying spaces and classifying topoi / Izak Moerdijk. -
Berlin ; Heidelberg ; New York ; Barcelona ; Budapest ; Hong
Kong ; London ; Milan ; Paris ; Tokyo : Springer, 1995
(Lecture notes in mathematics ; 1616)
ISBN 3-540-60319-0
NE: GT

Mathematics Subject Classification (1991): 18F10, 55N30, 55P15

ISBN 3-540-60319-0 Springer-Verlag Berlin Heidelberg New York

Printed in Germany

Typesetting: Camera-ready TEX output by the author
SPIN: 10479497 46/3142-543210 - Printed on acid-free paper

Preface

In these notes, a detailed account is presented of the relation between classifying spaces and classifying topoi. To make the notes more accessible, I have tried to keep the prerequisites to a minimum, for example by starting with an introductory chapter on topos theory, and by reviewing the necessary basic properties of geometric realization and classifying spaces in the first part of Chapter III. Furthermore, I have made an attempt to present the material in such a way that it is possible to read the special case of discrete categories first. This case already provides a good general picture, while it avoids some of the technical complications involved in the general case of topological categories. Thus, to reach the comparison and classification theorems for discrete categories in Section IV.1, the reader can omit §§3,4,5,7 and most of §6 in Chapter II, as well as the second parts of §1 and §2 in Chapter III.

In the past several years I have been helped by discussions with several people which were directly or indirectly related to the subject matter of these notes. In this respect, I am particularly indebted to W.T. van Est, S. Mac Lane, G. Segal and J.A. Svensson. Above all, A. Joyal taught me not to underestimate the Sierpinski space.

A summary of the main results appeared in the Comptes Rendus de l'Académie des Sciences (t. 317, 1993). The present version was mainly written during the fall of 1994, which I spent at the University of Aarhus. I am most grateful for the hospitality and support of the mathematical institute there. I would also like to thank A. Dold for the possibility to publish these notes in the Springer Lecture Notes Series, and for some advice on exposition. Finally, I would like to thank Elise Goeree for her careful typing of the manuscript.

This research is part of a project funded by the Dutch Organization for Scientific Research (NWO).

Utrecht, Spring 1995.

Contents

Introduction

These notes arose out of two related questions. First, what does the so-called classifying space of a small category actually classify? And secondly, what is the relation between classifying spaces and classifying topoi?

These questions can perhaps best be explained by describing the well-known case of a group G. The classifying space BG classifies principal G-bundles (or covering spaces with group G), in the sense that for any suitable space X (e.g., a CW-complex) there is a bijective correspondence between isomorphism classes of such covering projections $E \to X$ and homotopy classes of maps $X \to BG$. Furthermore, the cohomology groups of this space BG are exactly the Eilenberg-Mac Lane cohomology groups of the group G.

On the other hand, there is the classifying topos of the group G, introduced by Grothendieck and Verdier in SGA4, and defined as the category of all sets equipped with an action by the group G. I will denote this category by $\mathcal{B}G$. The topos $\mathcal{B}G$ has the same properties as the space BG, for tautological reasons: the cohomology of the topos $\mathcal{B}G$ is the group cohomology of G, because the definitions of topos cohomology and group cohomology are verbally the same in this case. And for any other topos $\mathcal{T}$, the fact that topos maps from $\mathcal{T}$ into $\mathcal{B}G$ correspond to principal G-bundles over $\mathcal{T}$ is an elementary consequence of the definition of a map between topoi.

To compare the classifying space BG and the classifying topos $\mathcal{B}G$ of G-sets, one first has to put these two objects in one and the same category. For this reason, we replace the space BG by its topos $Sh(BG)$ of all sheaves (of sets) on BG.

More generally, it will be explained in Chapter I how for any space X, the topos $Sh(X)$ of sheaves on X contains basically the same information as the space X itself, and should be viewed simply as the space X disguised as a topos. This view is supported by the fact that for two spaces X and Y, continuous mappings between X and Y correspond to topos mappings between $Sh(X)$ and $Sh(Y)$. Moreover, for a sufficiently good space X (e.g., a CW-complex), the cohomology groups of the space X are the same as those of the topos $Sh(X)$.

To come back to the comparison between the space BG and the topos $\mathcal{B}G$ of G-sets, we note that after having replaced BG by its topos $Sh(BG)$, the two can be related by a mapping $Sh(BG) \to \mathcal{B}G$. This topos map is a weak homotopy equivalence, although $\mathcal{B}G$ is a much smaller and simpler topos than $Sh(BG)$. The known isomorphisms between the cohomology and homotopy groups of the space BG and those of the topos $\mathcal{B}G$ are induced by this map $Sh(BG) \to \mathcal{B}G$. Furthermore, it follows that for a CW-complex X, there is a bijective correspondence between homotopy classes of maps between spaces $X \to BG$ and homotopy classes of topos maps $Sh(X) \to \mathcal{B}G$. In this way, the fact that the space BG classifies principal G-bundles can be seen as a consequence of the fact that the topos $\mathcal{B}G$ does.

The first purpose in these notes will be to extend this relation between classifying space and classifying topos from the well-known and elementary case of a group G to that of an arbitrary small category $\mathbb{C}$. In Chapter I, we will recall how the classifying topos $\mathcal{B}\mathbb{C}$ of $\mathbb{C}$ is constructed as the topos of all presheaves on $\mathbb{C}$, i.e. of all contravariant set-valued functors on $\mathbb{C}$. In Chapter III, §2, it will be recalled how the classifying space $B\mathbb{C}$ is constructed as the geometric realization of the nerve of $\mathbb{C}$. As for groups, it is known that the two constructions define the same cohomology (for locally constant, abelian coefficients). We will relate the two constructions, by first replacing the space $B\mathbb{C}$ by its topos $Sh(B\mathbb{C})$, and then constructing a weak homotopy equivalence of topoi (see Theorem 1.1 in Chapter IV):

$$p : Sh(B\mathbb{C}) \longrightarrow \mathcal{B}\mathbb{C}. \tag{1}$$

The construction of this map p is based on a comparison of various types of geometric realization, for spaces as well as for topoi and using different kinds of intervals, to be presented in Chapter III.

Of course, a lot more information is contained in a weak homotopy equivalence (1) than in the mere fact that the space $B\mathbb{C}$ and the topos $\mathcal{B}\mathbb{C}$ have isomorphic cohomology groups. For example, from the existence of such a map $Sh(B\mathbb{C}) \to \mathcal{B}\mathbb{C}$, one can conclude that for any CW-complex X there is a bijective correspondence between homotopy classes of maps of spaces $X \to B\mathbb{C}$ and homotopy classes of maps of topoi $Sh(X) \to \mathcal{B}\mathbb{C}$:

$$[X, B\mathbb{C}] = [Sh(X), \mathcal{B}\mathbb{C}]. \tag{2}$$

Using this bijective correspondence, one can transfer known classification results for the topos $\mathcal{B}\mathbb{C}$ to the space $B\mathbb{C}$. Indeed, define a principal $\mathbb{C}$-bundle E on a space X to be a system of sheaves $E(c)$, one for each object c in $\mathbb{C}$, on which $\mathbb{C}$ acts by sheaf maps $\alpha_* : E(c) \to E(d)$ for each arrow $\alpha : c \to d$ in $\mathbb{C}$, in a functorial way. Moreover, the bundle E should satisfy the following three conditions for being principal, for each point x in X (where $E(c)_x$ denotes the stalk of $E(c)$ at x):

(i) $\bigcup_{c \in \mathbb{C}} E(c)_x$ is non-empty.

(ii) The action is transitive: given $y \in E(c)_x$ and $z \in E(d)_x$, there are arrows $\alpha : b \to c$ and $\beta : b \to d$ in $\mathbb{C}$ and a point $w \in E(b)_x$ for which $\alpha_*(w) = y$ and $\beta_*(w) = z$.

(iii) The action is free: given $y \in E(c)_*$ and parallel arrows $\alpha, \beta : c \rightrightarrows d$ in $\mathbb{C}$ so that $\alpha_*(y) = \beta_*(y)$, there exists an arrow $\gamma : b \to c$ in $\mathbb{C}$ and a point $z \in E(b)_x$, for which $\alpha\gamma = \beta\gamma$ and $\gamma_*(z) = y$.

Note that in case $\mathbb{C}$ is a group (viewed as a category with only one object), this definition of principal bundle agrees with the usual one.

A basic result of topos theory, which we will review in Section II.2, states that there is an exact correspondence between such principal $\mathbb{C}$-bundles over X and topos maps $Sh(X) \to \mathcal{B}\mathbb{C}$. Using this correspondence and the bijection (2) above, one obtains for a CW-complex X and a small category $\mathbb{C}$ the following theorem, to be proved in Section IV.1:

Theorem. *Homotopy classes of maps $X \to B\mathbb{C}$ are in bijective correspondence with concordance classes of principal $\mathbb{C}$-bundles over X.*

Here two principal bundles over X are said to be concordant if they lie at the two ends of some principal bundle over $X \times [0,1]$.

This theorem of course contains the classical fact that the classifying space BG of a group G classifies principal G-bundles. The theorem also extends a result of G. Segal, which states that for a monoid with cancellation M, its classifying space BM classifies a suitably defined notion of principal M-bundle.

Thus, the weak equivalence (1) and the theorem above together provide an answer to the two questions stated at the beginning of this introduction, for the case of a discrete category $\mathbb{C}$.

Much of the work in these notes is concerned with the problem of extending these results to topological categories. Recall that a topological category $\mathbb{C}$ is given by a space of objects $\mathbb{C}_0$ and a space of arrows $\mathbb{C}_1$, together with continuous operations for source and target $\mathbb{C}_1 \rightrightarrows \mathbb{C}_0$, for identity arrows $\mathbb{C}_0 \to \mathbb{C}_1$, and for composition $\mathbb{C}_1 \times_{\mathbb{C}_0} \mathbb{C}_1 \to \mathbb{C}_1$. For example, any topological group or monoid is a topological category (with a space of objects which consists of just one point), as is any topological groupoid, such as the holonomy groupoid of a foliation (Haefliger(1984), Bott(1972), Segal(1968)). The construction of the classifying topos of a topological category will be described in detail in §II.3 while the classical construction of the classifying space will be reviewed in §III.2. The general considerations concerning geometric realization will again provide a map as in (1) relating the classifying space and the classifying topos. This map will in general not be a weak homotopy equivalence. However, there is an interesting case, which includes that of discrete categories, where the map is a weak homotopy equivalence. This is the case of topological categories $\mathbb{C}$ with the property that their source map $s : \mathbb{C}_1 \to \mathbb{C}_0$ is étale, i.e. is a local homeomorphism. Such topological categories will be called s-étale. For example, many of the topological groupoids arising in the theory of foliations are s-étale, as are topological categories constructed from diagrams of spaces (see §II.5 below). For s-étale topological categories, the map (1) is a weak homotopy equivalence, as said; moreover, it will be shown in §II.4 that the correspondence between topos maps into $\mathcal{B}\mathbb{C}$ and principal $\mathbb{C}$-bundles, already referred to above, generalizes to the case of s-étale topological categories $\mathbb{C}$. It follows that the theorem just stated also holds for s-étale topological categories.

As an illustration of the use of classifying topoi for discrete and s-étale categories, we will present in §IV.3 a relatively straightforward topos theoretic proof of Segal's theorem on the weak homotopy type of the Haefliger groupoid Γ^q.

For an arbitrary topological groupoid (not necessarily s-étale) the "naively" constructed classifying topos $\mathcal{B}\mathbb{C}$ need not contain much information. To obtain a suitable comparison with the classifying space $B\mathbb{C}$, we will consider a different classifying topos for $\mathbb{C}$, described by Deligne. Recall that in Deligne(1975), the notion of a sheaf on a simplicial space Y is introduced, and the topos $Sh(Y)$ (Deligne writes $\tilde{Y}$) of all such sheaves is considered as an alternative for the geometric realization $|Y|$. In particular for a topological category $\mathbb{C}$, and its associated simplicial space Nerve($\mathbb{C}$), the topos Sh(Nerve($\mathbb{C}$)) provides an alternating for the classifying space $B\mathbb{C}$. This topos Sh(Nerve($\mathbb{C}$)) will be called the Deligne classifying topos of $\mathbb{C}$, and be denoted by $\mathcal{D}\mathbb{C}$.

Deligne shows in op. cit. that for a suitable simplicial space Y the realization and the topos of sheaves $Sh(Y)$ have isomorphic cohomology groups. In §IV.4 it will be shown that this isomorphism in cohomology is induced by a map, and that the topos $Sh(Y)$ has the same weak homotopy type as the geometric realization $|Y|$. In particular, this will show that for any topological category $\mathbb{C}$, the Deligne classifying topos $\mathcal{D}\mathbb{C}$ and the classifying space $B\mathbb{C}$ have the same weak homotopy type. From this last result, one can obtain an answer to the question what $B\mathbb{C}$ classifies: it will be shown that homotopy classes of maps $X \to B\mathbb{C}$ correspond to concordance classes of sheaves of linear orders on X equipped with a suitable augmentation into the category $\mathbb{C}$.

These notes by no means provide a complete picture of the comparison between classifying spaces and classifying topoi for topological categories, and many questions remain. One obvious question for the case of a topological group(oid) G is the precise relation between linear orders augmented by G which are shown to be classified by BG in these notes, and principal G-bundles. Another question concerns the relationship between the "small" classifying topoi $\mathcal{B}\mathbb{C}$ and $\mathcal{D}\mathbb{C}$ of a topological category, and the classifying "gros" topoi defined over the topological gros topos by Grothendieck and Verdier (see e.g. SGA4 (tome 1), p.317)).

Chapter I

Background in Topos Theory

§1 Basic definitions

A topos is a "generalized" topological space. Indeed according to Grothendieck, topoi (should) form the proper subject of study for topology. The basic idea is similar to that of various well-known dualities. For example, Gelfand duality states that one could replace a compact Hausdorff space X by its ring $C(X)$ of complex-valued functions; mappings between such spaces can be described in terms of these rings, and the space X can be recovered (up to homeomorphism) from $C(X)$.

Similarly, one can use the "ring" (category) of sets instead of the ring of complex numbers, and replace a space X by the collection of all its "continuous set-valued functions"; i.e. the sheaves of sets on X, described in detail in the next section. As for Gelfand duality, mappings between spaces can be described in terms of these sheaves, and the space X can be recovered from the collection of all its sheaves.

The definition of a topos is meant to capture the basic properties of this category of all sheaves on a space X, and similar categories. Sheaves of sets are taken as basic here, since abelian sheaves, simplicial sheaves, etc., can all be defined in terms of sheaves of sets. We present the definition of a topos in the "Giraud form", which requires some elementary categorical notions to be explained first. (For background in category theory, Chapters I - IV of Mac Lane(1971) suffice.)

Let $\mathcal{E}$ be a category. (It is our convention that the objects of $\mathcal{E}$ can form a proper class, but that for any two objects A and B the collection $\mathrm{Hom}(A, B)$ of all arrows from A to B is a set. If the objects of $\mathcal{E}$ form a set as well, $\mathcal{E}$ is said to be "small".)

1.1. Definition. A category $\mathcal{E}$ is said to be a *topos* iff it satisfies the Giraud axioms (G1-G4), to be stated below.

(G1) The category $\mathcal{E}$ has finite limits.

This axiom needs no further explanation. For the second axiom, we recall that a sum (coproduct) $\sum_{i \in I} E_i$, indexed by some set I, is said to be *disjoint* if for any two distinct indices j and k the diagram

$$\begin{array}{ccc} 0 & \longrightarrow & E_k \\ \downarrow & & \downarrow \\ E_j & \longrightarrow & \sum E_i \end{array}$$

is a pullback; here the maps into the coproduct are the canonical ones, and 0 denotes the initial object of $\mathcal{E}$ (this is the sum of the empty family). If each E_i is equipped with an arrow $E_i \to A$ into a given object A, then the sum also has such an evident arrow $\sum E_i \to A$. Thus for any map $B \to A$, there is a canonical map

$$\sum B \times_A E_i \;\to\; B \times_A \sum E_i \,. \tag{1}$$

Sums in $\mathcal{E}$ are said to be *stable* if this map (1) is always an isomorphism – in other words, if sums commute with pullbacks. The second Giraud axiom now is:

(G2) All (set-indexed) sums exist in $\mathcal{E}$, and are disjoint and stable.

For the next axiom, consider an object E in $\mathcal{E}$ and a monomorphism $r : R \rightarrowtail E \times E$. For any object T in $\mathcal{E}$, composition with r defines a subset

$$\mathrm{Hom}(T, R) \subseteq \mathrm{Hom}(T, E \times E) \cong \mathrm{Hom}(T, E) \times \mathrm{Hom}(T, E).$$

If, for every object T, this subset is an equivalence relation on the set $\mathrm{Hom}(T, E)$, then the monomorphism $r : R \rightarrowtail E \times E$ is said to be an *equivalence relation* on E. For example, if $f : E \to F$ is any arrow, then the pullback $E \times_F E \rightarrowtail E \times E$ is an equivalence relation on E. A diagram

$$R \underset{r_2}{\overset{r_1}{\rightrightarrows}} E \xrightarrow{f} F$$

in $\mathcal{E}$ is said to be *exact* if f is the coequalizer of r_1 and r_2 and

$$\begin{array}{ccc} R & \xrightarrow{r_2} & E \\ {\scriptstyle r_1}\downarrow & & \downarrow{\scriptstyle f} \\ E & \xrightarrow[f]{} & F \end{array}$$

is a pullback. It is said to be *stably exact* if for any arrows $F \to A \leftarrow B$, the diagram

$$B \times_A R \;\rightrightarrows\; B \times_A E \;\to\; B \times_A F \,,$$

obtained by pullback along $B \to A$, is again exact. The third Giraud axiom is:

(G3) (a) For every epimorphism $E \to F$ in $\mathcal{E}$, the diagram $E \times_F E \rightrightarrows E \to F$ is stably exact.

(b) For every equivalence relation $R \rightarrowtail E \times E$, there exists an object E/R which fits into an exact diagram $R \rightrightarrows E \to E/R$.

It follows that any exact diagram in $\mathcal{E}$ is stably exact. It also follows that all small colimits exist in the category $\mathcal{E}$, since these can be constructed from sums and coequalizers of equivalence relations (as in G3 (b)); see Mac Lane - Moerdijk (1992), p. 577.

For the last axiom, recall that a collection of objects $\{G_i : i \in I\}$ of $\mathcal{E}$ is said to *generate* $\mathcal{E}$ when for any two parallel arrows $u, v : E \rightrightarrows F$ in $\mathcal{E}$, if $u \circ t = v \circ t$ for every arrow $t : G_i \to E$ from every G_i, then $u = v$. The collection $\{G_i : i \in I\}$ of objects is said to be *small* if it is a set (rather than a proper class).

(G4) The category $\mathcal{E}$ has a small collection of generators.

If $\{G_i : i \in I\}$ is a set of generators, then every object E in $\mathcal{E}$ is a colimit of such generating objects.

A *morphism* between topoi $f : \mathcal{F} \to \mathcal{E}$ consists of a pair of functors ("inverse" and "direct" image functors)

$$f^* : \mathcal{E} \to \mathcal{F} \text{ and } f_* : \mathcal{F} \to \mathcal{E}$$

with the following two properties:

(i) f^* is left adjoint to f_* ; i.e. there is a natural isomorphism

$$\mathrm{Hom}_{\mathcal{F}}(f^*E, F) \cong \mathrm{Hom}_{\mathcal{E}}(E, f_*F),$$

(ii) f^* commutes with finite limits (i.e., is "left exact").

Such morphisms $f : \mathcal{F} \to \mathcal{E}$ and $g : \mathcal{G} \to \mathcal{F}$ can be composed in the evident way,

$$(f \circ g)^* = g^* \circ f^* \ , \ (f \circ g)_* = f_* \circ g_* \, .$$

Since the inverse image f^* of any morphism f is a left adjoint, it commutes with colimits. Therefore, since every object of $\mathcal{E}$ is a colimit of generators, f^* is completely determined (up to natural isomorphism) by its behaviour on generators. Furthermore, any functor $f^* : \mathcal{E} \to \mathcal{F}$ which commutes with colimits must have a right adjoint, necessarily unique up to isomorphism (Mac Lane (1971), p. 83). Thus topos morphisms can be described more economically, and we see will some explicit examples of this later.

The collection of all morphisms $f : \mathcal{F} \to \mathcal{E}$ has itself the structure of a category: an arrow δ between two morphisms $f, g : \mathcal{F} \to \mathcal{E}$ is a natural transformation

$$\delta : f^* \to g^*$$

between the inverse image functors. By the remarks above, this category $\mathrm{Hom}(\mathcal{F}, \mathcal{E})$ is equivalent to the category of functors $f^* : \mathcal{E} \to \mathcal{F}$ which commute with colimits and finite limits, and natural transformations between them.

A morphism $f : \mathcal{F} \to \mathcal{E}$ is said to be an *equivalence* if there exists a morphism $g : \mathcal{E} \to \mathcal{F}$ and isomorphisms $f \circ g \cong \mathrm{id}_{\mathcal{E}}$ and $g \circ f \cong \mathrm{id}_{\mathcal{F}}$. This is equivalent to the requirement that the unit $id_{\mathcal{E}} \to f_*f^*$ and the counit $f^*f_* \to id_{\mathcal{F}}$ are natural isomorphisms. The topoi $\mathcal{E}$ and $\mathcal{F}$ are said to be equivalent if there exists such an equivalence f, and one writes

$$\mathcal{E} \cong \mathcal{F}$$

in this case. In practice, one often tacitly identifies equivalent topoi, just as one identifies homeomorphic spaces. However, given two topoi $\mathcal{E}$ and $\mathcal{F}$, one cannot always identify isomorphic morphisms $\mathcal{F} \to \mathcal{E}$, as will be clear, e.g. from the discussion of pushouts of topoi in Section 3.

If $\mathcal{E}$ is a topos and B is an object in $\mathcal{E}$, one can form the "comma-category" $\mathcal{E}/B$, with as objects the arrows $E \to B$ in $\mathcal{E}$, and as arrows in $\mathcal{E}/B$ the commutive triangles in $\mathcal{E}$. This category $\mathcal{E}/B$ again satisfies the Giraud axioms for a topos: it inherits all the required exactness properties from $\mathcal{E}$; and if $\{G_i : i \in I\}$ is a set of generators for $\mathcal{E}$, then the collection of all arrows $G_i \to B$ (for all $i \in I$) is a set of generators for $\mathcal{E}$. The functor $E \mapsto (\pi_2 : E \times B \to B) : \mathcal{E} \to \mathcal{E}/B$ commutes with colimits and finite limits, and hence is the inverse image functor of a topos morphism $\mathcal{E}/B \to \mathcal{E}$. (Its direct image part Π_B is described explicitly, e.g. in Mac Lane-Moerdijk (1992),p. 60.)

§2 First examples

In this section we describe the topos of sheaves on a space and the topos of presheaves on a small category. Before doing so, we should mention the simplest example of a topos, viz. the category of all small sets, denoted $\mathcal{S}$ or (*sets*). (One readily verifies the Giraud axioms (G1-4) for $\mathcal{S}$; for a collection of generators, one can take the one-element collection consisting of the one-point set.)

For any other topos $\mathcal{E}$, there is a morphism $\gamma : \mathcal{E} \to \mathcal{S}$, unique up to isomorphism. It can be described explicitly, in terms of the terminal object 1 of $\mathcal{E}$, by

$$\gamma^*(S) = \sum_{s \in S} 1 \ , \ \ \gamma_*(E) = \mathrm{Hom}_{\mathcal{E}}(1, E) \ ,$$

for any set S and any object E in $\mathcal{E}$. One often writes Δ for γ^* and Γ for γ_*. The functor Δ is called the *constant sheaf functor*, and Γ the *global sections functor*.

Now let X be a topological space. A continuous map $f : E \to X$ is said to be a *local homeomorphism* (or, an *étale map*, or an *étale space* over X) if both f and its diagonal $E \to E \times_X E$ are open maps. This is equivalent to the requirement that for any point $y \in E$ there exist open neighbourhoods $V_y \subseteq E$ and $U_{f(y)} \subseteq X$ so that f restricts to a homeomorphism $f : V_y \xrightarrow{\sim} U_{f(y)}$. A *sheaf on* X is such an étale map $f : E \to X$. A map φ between sheaves $(f : E \to X) \to (f' : E' \to X)$ is a continuous map $\varphi : E \to E'$ so that $f' \circ \varphi = f$. This defines a category of all sheaves on X, denoted

$$Sh(X) .$$

[In the literature, one often defines a sheaf (of sets) as a functor $F : \mathcal{O}(X)^{op} \to (sets)$, defined on the poset $\mathcal{O}(X)$ of all open subsets of X, and having for each open cover $U = \bigcup U_i$ the "unique pasting property" that the diagram

$$F(U) \to \prod_i F(U_i) \rightrightarrows \prod_{i,j} F(U_i \cap U_j)$$

is an equalizer of sets. These definitions are of course equivalent, as is explained in any book on sheaf theory; see e.g. Godement(1958), Swan(1964).]

The category $Sh(X)$ is a topos. Indeed, finite limits and colimits are constructed just as for topological spaces, because these constructions preserve étale maps. More explicitly, if $E \to X$ and $F \to X$ are two étale maps then so are $E \times_X F \to X$ and $E + F \to X$, and these represent the product and sum in the category $Sh(X)$. The same applies to infinite sums. Similarly, in an exact diagram of topological spaces over X,

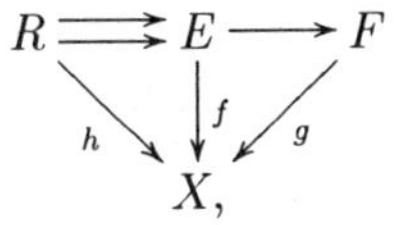

if f and g are étale then so is h, while if h and f are étale then so is g. Thus $Sh(X)$ inherits all the relevant exactness properties from topological spaces. For the set of generators, one can take the collection of all embeddings $U \hookrightarrow X$ of open subsets of X. To see that these generate, take two distinct parallel maps a and b between sheaves

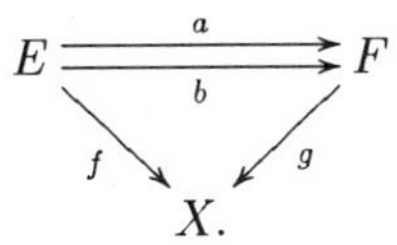

Let $e \in E$ be a point with $a(e) \neq b(e)$, and let V_e be a small neighbourhood of e so that $f : V_e \to f(V_e) = U$ is a homeomorphism onto the open set U. Then f^{-1} defines a map of sheaves from $(U \hookrightarrow X)$ to $(E \to X)$ with the property that $a \circ f^{-1} \neq b \circ f^{-1}$.

From the topos $Sh(X)$ of sheaves on X, one can recover the lattice $\mathcal{O}(X)$ of open subsets of X, essentially as the subcategory consisting of all sheaves $(E \to X)$ with the property that the unique map into the terminal object $1 = (id : X \to X)$ of $Sh(X)$ is a monomorphism. Thus we can recover the space X from $Sh(X)$ provided the points of X are determined by their open neighbourhoods. This is the case precisely when the space X is *sober*. [Recall, from SGA IV, vol. 1, p. 336, that a closed set F in X is irreducible if it cannot be written as the union of two smaller closed sets, and that X is sober if *every* such irreducible closed set F is of the from $F = \overline{\{x\}}$ for a unique point x. Every Hausdorff space is sober.] *In this text, all spaces will be assumed to be sober.*

A continuous map $f : Y \to X$ between spaces induces two well-known adjoint functors

$$f^* : Sh(X) \to Sh(Y) \ , \ f_* : Sh(Y) \to Sh(X)$$

between the categories of sheaves of sets. In terms of étale spaces, f^* is simply pullback (fibered product) along f. It evidently preserves (finite) limits and colimits. The right adjoint f_* is more easily described in terms of sheaves as functors: for a sheaf $F : \mathcal{O}(Y)^{op} \to (sets)$,

$$f_*(F) = F \circ f^{-1} : \mathcal{O}(X)^{op} \to \mathcal{O}(Y)^{op} \to (sets) \ .$$

These two functors constitute a morphism of topoi, denoted

$$f : Sh(Y) \to Sh(X) \ .$$

Conversely, suppose $\varphi : Sh(Y) \to Sh(X)$ is any morphism of topoi. Then the functor φ^*, when restricted to subobjects of the terminal object, gives an operation $\varphi^* : \mathcal{O}(X) \to \mathcal{O}(Y)$ which preserves finite intersections and arbitrary unions. For a point $y \in Y$, define $F_y = X - \bigcup\{U \in \mathcal{O}(X) : y \notin \varphi^*(U)\}$. Then F_y is an irreducible closed set, so if Y is sober there is a unique point $x = \varphi(y)$ so that $F_y = \overline{\{x\}}$. This defines a map $\varphi : Y \to X$ with the property that for any open set $U \subseteq X$, and any point $y \in Y$, $\varphi(y) \in U$ iff $y \in \varphi^*(U)$. In this way, the map $\varphi : Y \to X$ is determined by the inverse image functor φ^*.

For sober spaces X and Y, these constructions set up a correspondence between continuous maps $Y \to X$ and (isomorphism classes of) topos morphisms $Sh(Y) \to Sh(X)$. Thus, the assignment

$$X \mapsto Sh(X)$$

of the topos of sheaves to a sober space X doesn't change the notion of mapping, and the topos $Sh(X)$ should simply be viewed as a faithful image of the space X in the world of topoi. Indeed, we will in the sequel often simply write X when it is evident that we mean the topos of sheaves on the space X. For example, when $\mathcal{E}$ is another topos, an arrow

$$X \to \mathcal{E}$$

denotes a topos morphism $Sh(X) \to \mathcal{E}$. In Section 4 below, we will discuss how algebraic invariants of the space X such as homotopy and cohomology groups can be defined in terms of the topos $Sh(X)$.

For the second elementary example of a topos, consider a small category $\mathbf{C}$. A *presheaf* (of sets) on $\mathbf{C}$ is a functor

$$S : \mathbf{C}^{op} \to (sets) \ .$$

Thus S assigns to each object $x \in \mathbf{C}$ a set $S(x)$, and to each arrow $\alpha : x \to y$ in $\mathbf{C}$ a function $S(\alpha) : S(y) \to S(x)$, called restriction along α and denoted

$$s \mapsto s \cdot \alpha = S(\alpha)(s) \quad (\text{for } s \in S(y)).$$

The functoriality of S is then reflected in the usual identities $s \cdot 1 = s$ and $(s \cdot \alpha) \cdot \beta = s \cdot (\alpha\beta)$ for an action. As morphisms $\varphi : S \to T$ between two such presheaves we take the natural transformations. Thus φ is given by functions $\varphi_x : S(x) \to T(x)$ (for each object x in $\mathbf{C}$), which respect the restrictions:

$$\varphi_x(s \cdot \alpha) = \varphi_y(s) \cdot \alpha ,$$

for s and α as above. This category of all presheaves on $\mathbf{C}$ is denoted as a functor category $sets^{\mathbf{C}^{op}}$, or as

$$\mathcal{B}\mathbf{C} .$$

This category $\mathcal{B}\mathbf{C}$ is a topos, called the *classifying topos* of the category $\mathbf{C}$. To see that the Giraud axioms are satisfied, note first that all limits and colimits of presheaves can be constructed "pointwise", as in

$$(\varprojlim S_i)(x) = \varprojlim S_i(x) \ , \ (\varinjlim S_i)(x) = \varinjlim S_i(x) .$$

Therefore all limits and colimits of presheaves, in particular pullbacks, sums and coequalizers, inherit all exactness properties from the category of sets. Thus it is clear that $\mathcal{B}\mathbf{C}$ satisfies the Giraud axioms (G1)-(G3).

For the axiom (G4) on generators, consider the "Yoneda embedding" $\mathrm{Yon} : \mathbf{C} \hookrightarrow \mathcal{B}\mathbf{C}$, defined by

$$\mathrm{Yon}(x)(y) = \mathrm{Hom}_{\mathbf{C}}(y, x) .$$

Thus $\mathrm{Yon}(x)$ is the *representable presheaf* given by x. The so-called Yoneda lemma states that for any presheaf S, there is a natural isomorphism

$$\theta = \theta_S : \mathrm{Hom}_{\mathcal{B}\mathbf{C}}(\mathrm{Yon}(x), S) \cong S(x) , \qquad (1)$$

defined for a natural transformation $\varphi : \mathrm{Yon}(x) \to S$ by

$$\theta(\varphi) = \varphi_x\,(id_x) .$$

Naturality of θ means that for any morphism $\psi : S \to T$ of presheaves, the diagram

$$\begin{array}{ccc} \mathrm{Hom}_{\mathcal{B}\mathbf{C}}(\mathrm{Yon}(x), S) & \xrightarrow{\theta_S} & S(x) \\ \downarrow{\scriptstyle \psi_*} & & \downarrow{\scriptstyle \psi_x} \\ \mathrm{Hom}_{\mathcal{B}\mathbf{C}}(\mathrm{Yon}(x), T) & \xrightarrow{\theta_T} & T(x) \end{array}$$

commutes, where ψ_* denotes "composition with ψ". In particular, ψ is completely determined by all composites $\mathrm{Yon}(x) \xrightarrow{\varphi} S \xrightarrow{\psi} T$ from representable presheaves $\mathrm{Yon}(x)$. Thus, the collection of these presheaves, for all $x \in \mathbf{C}$, generates $\mathcal{B}\mathbf{C}$. This proves that $\mathcal{B}\mathbf{C}$ satisfies axiom (G4).

A functor $f : \mathbf{D} \to \mathbf{C}$ between small categories induces an evident operation f^* on presheaves, by composition:

$$f^* : \mathcal{B}\mathbf{C} \to \mathcal{B}\mathbf{D} \ , \ f^*(S)(y) = S(fy) .$$

This functor f^* evidently preserves limits and colimits, since these are all computed pointwise. Furthermore, f^* has a right adjoint $f_* : \mathcal{B}\mathbf{D} \to \mathcal{B}\mathbf{C}$, defined by

$$f_*(T)(x) = \mathrm{Hom}_{\mathcal{B}\mathbf{D}}(f^*(\mathrm{Yon}(x)), T) .$$

The adjunction isomorphism

$$\mathrm{Hom}_{\mathcal{B}\mathbf{D}}(f^*(S), T) \cong \mathrm{Hom}_{\mathcal{B}\mathbf{C}}(S, f_*(T))$$

can be described as follows: given $\varphi : f^*(S) \to T$, construct $\bar{\varphi} : S \to f_*(T)$ with components $\bar{\varphi}_x : S(x) \to f_*(T)(x)$ defined via the isomorphism θ in (1) as

$$\begin{array}{ccccc}
S(x) & & \xrightarrow{\quad\bar{\varphi}_x\quad} & & f_*(T)(x) \\
\downarrow{\scriptstyle \theta^{-1}} & & & & \| \\
\mathrm{Hom}_{\mathcal{B}\mathbf{C}}(\mathrm{Yon}(x), S) & \xrightarrow{f^*} & \mathrm{Hom}_{\mathcal{B}\mathbf{D}}(f^*\mathrm{Yon}(x), f^*S) & \xrightarrow{\varphi_*} & \mathrm{Hom}_{\mathcal{B}\mathbf{D}}(f^*\mathrm{Yon}(x), T)
\end{array}$$

where φ_* denotes composition with φ. Conversely, given $\psi : S \to f_*(T)$, construct $\tilde{\psi} : f^*(S) \to T$ with components $\tilde{\psi}_y : f^*(S)(y) = S(fy) \to T(y)$, using the evident map $\mathrm{Yon}(y) \to f^*(\mathrm{Yon}(fy))$, as

$$\begin{array}{ccccc}
S(fy) & & \xrightarrow{\quad\tilde{\psi}_y\quad} & & T(y) \\
\downarrow{\scriptstyle \psi_{fy}} & & & & \uparrow{\scriptstyle \theta^{-1}} \\
f_*(T)(fy) & = & \mathrm{Hom}(f^*(\mathrm{Yon}(fy)), T) & \longrightarrow & \mathrm{Hom}(\mathrm{Yon}(y), T).
\end{array}$$

Thus the functor $f : \mathbf{D} \to \mathbf{C}$ induces a morphism of topoi, (again) denoted

$$f : \mathcal{B}\mathbf{D} \to \mathcal{B}\mathbf{C} ,$$

given by these adjoint functors f^* and f_*.

This construction of a topos morphism $\mathcal{B}\mathbf{D} \to \mathcal{B}\mathbf{C}$ from a functor $\mathbf{D} \to \mathbf{C}$ extends to natural transformations. Indeed, a transformation $\tau : g \to f$ between two functors $f, g : \mathbf{D} \rightrightarrows \mathbf{C}$ induces another transformation

$$\tilde{\tau} : f^* \to g^* : \mathcal{B}\mathbf{C} \rightrightarrows \mathcal{B}\mathbf{D} ,$$

defined for a presheaf S on $\mathbf{C}$ and an object y in $\mathbf{D}$ by

$$(\tilde{\tau}_S)_y : f^*(S)(y) = S(fy) \stackrel{S(\tau_y)}{\to} S(gy) = g^*(S)(y) .$$

Unlike the case of (sober) topological spaces, it is not true that all topos morphisms $\mathcal{B}\mathbf{D} \to \mathcal{B}\mathbf{C}$ come from functors $\mathbf{D} \to \mathbf{C}$. Indeed, there are many more morphisms $\mathcal{B}\mathbf{D} \to \mathcal{B}\mathbf{C}$ then there are functors $\mathbf{D} \to \mathbf{C}$, as will be evident from Chapter II, Section 2. In general, one cannot reconstruct the category $\mathbf{C}$ from the presheaf topos $\mathcal{B}\mathbf{C}$ either, because the representable presheaves are not characterized by a purely categorical property. (The closest one gets is by considering the class of all projective and connected presheaves: These are exactly the retracts of representable presheaves. If all idempotents split in $\mathbf{C}$, then every such retract is itself representable, up to isomorphism.)

§3 Some constructions of topoi

In this section we will describe the "universal" constructions of topoi, such as (fibered) products, amalgamated sums (pushouts) and inductive limits, to be used in later chapters.

Colimits of topoi all exist (Moerdijk(1988)), and are generally quite easy to describe. For example, for two topoi $\mathcal{E}$ and $\mathcal{F}$ their sum $\mathcal{E}+\mathcal{F}$ is by definition the topos for which there exists an equivalence of categories

$$\mathrm{Hom}(\mathcal{E},\mathcal{G}) \times \mathrm{Hom}(\mathcal{F},\mathcal{G}) \xrightarrow{\sim} \mathrm{Hom}(\mathcal{E} + \mathcal{F},\mathcal{G}) , \tag{1}$$

for any topos $\mathcal{G}$, and natural in $\mathcal{G}$. This sum $\mathcal{E} + \mathcal{F}$ can simply be constructed as the "categorical product": objects of $\mathcal{E} + \mathcal{F}$ are pairs (E, F) where E is an object of $\mathcal{E}$ and F one of $\mathcal{F}$, while arrows $(E,F) \to (E',F')$ in $\mathcal{E}+\mathcal{F}$ are pairs of arrows $E \to E'$ in $\mathcal{E}$ and $F \to F'$ in $\mathcal{F}$. It is easy to see that this category of pairs again satisfies the Giraud axioms for a topos. The equivalence (1) associates to a pair of morphisms $f : \mathcal{E} \to \mathcal{G}$ and $g : \mathcal{F} \to \mathcal{G}$ the unique (up to isomorphism) $h : \mathcal{E} + \mathcal{F} \to \mathcal{G}$ with $h^*(G) = (f^*E, g^*E)$.

Thus the sum of topoi is constructed as the product of categories. Note that for the two examples in the previous section, this corresponds to the usual sum of topological spaces and small categories: since a sheaf on the disjoint sum of spaces is the same thing as a pair of sheaves, one has

$$Sh(X+Y) \cong Sh(X) + Sh(Y) .$$

Similarly, for small categories $\mathbb{C}$ and $\mathbb{D}$,

$$\mathcal{B}(\mathbb{C} + \mathbb{D}) \cong \mathcal{B}(\mathbb{C}) + \mathcal{B}(\mathbb{D}) .$$

For two morphisms of topoi $f : \mathcal{E} \to \mathcal{F}$ and $g : \mathcal{E} \to \mathcal{G}$, their pushout (amalgamated sum) $\mathcal{F} \cup_{\mathcal{E}} \mathcal{G}$ is described as follows. There is a square

$$\begin{array}{ccc} \mathcal{E} & \xrightarrow{g} & \mathcal{G} \\ {\scriptstyle f}\downarrow & \overset{\alpha}{\nearrow} & \downarrow{\scriptstyle v} \\ \mathcal{F} & \xrightarrow[u]{} & \mathcal{F} \cup_{\mathcal{E}} \mathcal{G} \end{array} \tag{2}$$

which commutes up to a given isomorphism $\alpha : uf \cong vg$, and with the following universal property: for any topos $\mathcal{H}$, the functor from the category $\mathrm{Hom}(\mathcal{F} \cup_{\mathcal{E}} \mathcal{G},\mathcal{H})$ to the category of triples

$$(\varphi : \mathcal{F} \to \mathcal{H} \, , \; \psi : \mathcal{G} \to \mathcal{H} \, , \; \beta : \varphi f \cong \psi g) ,$$

which sends a map $h : \mathcal{F} \cup_{\mathcal{E}} \mathcal{G} \to \mathcal{H}$ to $(hu, hv, h \circ \alpha)$, is an equivalence of categories. This universal property determines $\mathcal{F} \cup_{\mathcal{E}} \mathcal{G}$ uniquely, up to equivalence of topoi.

Analogous to the case of sums, this pushout of topoi can be constructed explicitly as a fibered product of categories: $\mathcal{F} \cup_{\mathcal{E}} \mathcal{G}$ is the category with as objects triples

(F, G, a) where F is an object of $\mathcal{F}$ and G one of $\mathcal{G}$, while $a : f^*F \xrightarrow{\sim} g^*G$ is an isomorphism in $\mathcal{E}$. Arrows $(F, G, a) \to (F', G', a')$ in $\mathcal{F} \cup_{\mathcal{E}} \mathcal{G}$ are pairs of arrows $b : F \to F'$ in $\mathcal{F}$ and $c : G \to G'$ in $\mathcal{G}$, so that $a' \circ f^*(b) = g^*(c) \circ a$. Colimits and finite limits in this category of triples $\mathcal{F} \cup_{\mathcal{E}} \mathcal{G}$ can be constructed in the evident way from those in $\mathcal{F}$ and $\mathcal{G}$ (since f^* and g^* commute with colimits and finite limits), and one readily verifies that the Giraud axioms for a topos hold for $\mathcal{F} \cup_{\mathcal{E}} \mathcal{G}$. The morphisms u and v, required for the square (2), are defined by the evident inverse image functors

$$u^*(F, G, a) = F \quad , \quad v^*(F, G, a) = G ,$$

while $\alpha : uf \cong vg$ in (2) is the natural isomorphism with components

$$\alpha_{(F,G,a)} = a : (uf)^*(F, G, a) = f^*F \to g^*G = (vg)^*(F, G, a) .$$

For maps $X \leftarrow A \to B$ of topological spaces, there is a canonical morphism, comparing the pushout of topoi with that of spaces:

$$Sh(X) \cup_{Sh(A)} Sh(B) \to Sh(X \cup_A B) .$$

In Section 4 of Chapter III we will prove (and use) that this morphism is an equivalence of topoi for a closed embedding $A \hookrightarrow X$ between paracompact spaces.

Finally, we will use inductive limits (colimits) of sequences of topoi. For such a sequence

$$\mathcal{E}_o \xrightarrow{f_1} \mathcal{E}_1 \xrightarrow{f_2} \mathcal{E}_2 \to \cdots ,$$

the colimit $\mathcal{E}_\infty = \varinjlim \mathcal{E}_n$ is a topos equipped with morphisms $v_n : \mathcal{E}_n \to \mathcal{E}_\infty$ and isomorphisms $\alpha_n : v_n f_n \xrightarrow{\sim} v_{n-1}$, all together with the following universal property: For any topos $\mathcal{H}$, the evident functor from the category $\mathrm{Hom}(\mathcal{E}_\infty, \mathcal{H})$ to the category of pairs (u, β) where $u = (u_n)$ is a sequence of morphisms $u_n : \mathcal{E}_n \to \mathcal{H}$ and $\beta_n : u_n f_n \xrightarrow{\sim} u_{n-1}$, is an equivalence of categories.

Again, this inductive limit of topoi can be constructed as an inverse limit of categories. Define $\mathcal{E}_\infty$ to have as objects all pairs (E, a), where $E = (E_n)$ is a sequence of objects E_n in $\mathcal{E}_n$ while a is a sequence of isomorphisms $a_n : f_n^*(E_n) \xrightarrow{\sim} E_{n-1}$. The arrows $b : (E, a) \to (E', a')$ in $\mathcal{E}_\infty$ are sequences of arrows $b_n : E_n \to E'_n$ in $\mathcal{E}_n$, compatible with the a and a' in the sense that $b_{n-1} \circ a_n = a'_n \circ f_n^*(b_n)$. This category $\mathcal{E}_\infty$ is a topos, in which the finite limits and colimits are constructed in the evident way from those in each topos $\mathcal{E}_n$. For the required universal property, the morphisms $v_n : \mathcal{E}_n \to \mathcal{E}_\infty$ are given by the evident inverse image functors

$$v_n^*(E, a) = E_n ,$$

while the natural isomorphism $\alpha_n : v_n f_n \xrightarrow{\sim} v_{n-1}$ has components

$$(\alpha_n)_{(E,a)} = a_n : (v_n f_n)^*(E, a) = f_n^*(E_n) \to E_{n-1} = v_{n-1}^*(E, a) .$$

Later on, we will show and use that for a sequence of closed subspaces $X_0 \subseteq X_1 \subseteq \cdots$ of a paracompact space $X = \bigcup_n X_n$, the canonical comparison map

$$\varinjlim Sh(X_n) \longrightarrow Sh(\varinjlim X_n)$$

is an equivalence of topoi (cf. Chapter III, Section 4).

We will also use some products of topoi. For two topoi $\mathcal{E}$ and $\mathcal{F}$, their product $\mathcal{E} \times \mathcal{F}$ is the topos with the property that for any other topos $\mathcal{G}$, there is an equivalence of categories

$$\mathrm{Hom}(\mathcal{G}, \mathcal{E}) \times \mathrm{Hom}(\mathcal{G}, \mathcal{F}) \xrightarrow{\sim} \mathrm{Hom}(\mathcal{G}, \mathcal{E} \times \mathcal{F}),$$

natural in $\mathcal{G}$. This property determines $\mathcal{E} \times \mathcal{F}$ uniquely (up to equivalence of topoi). For two topoi $\mathcal{E}$ and $\mathcal{F}$, such a product $\mathcal{E} \times \mathcal{F}$ always exists, and is most easily constructed explicitly in terms of sites (cf. Mac Lane - Moerdijk, Chapter VII, Exercise 15). We will not need such an explicit description. The only property we will use is that for two topological spaces X and Y the canonical comparison map

$$Sh(X \times Y) \rightarrow Sh(X) \times Sh(Y) \qquad (3)$$

is an equivalence of topoi, whenever at least one of X, Y is locally compact. [For a proof, combine the fact that the functor which associates to a locale its topos of sheaves commutes with all products (see e.g. Joyal-Tierney(1984)) with the result that the product of two spaces agrees with their product as locales if one of the spaces is locally compact (see Dowker-Strauss(1977) and Isbell(1981)).]

§4 Cohomology and homotopy

In this section we review the standard definition for the cohomology and homotopy groups of a topos. Common references include SGA4 (vol. 2), Milne(1980), Artin-Mazur(1969).

Let $\mathcal{E}$ be a topos. We write $Ab(\mathcal{E})$ for the abelian category of abelian group objects in $\mathcal{E}$. For example, for the topos $Sh(X)$ of sheaves on X, this category $Ab(Sh(X))$ is the familiar category of abelian group valued sheaves. And for the topos $\mathcal{B}\mathbb{C}$ of presheaves on small category, $Ab(\mathcal{B}\mathbb{C})$ is the category of abelian presheaves, i.e. contravariant functors from $\mathbb{C}$ into the category Ab of abelian groups. The Giraud axiom (G4) for generators implies that the abelian category $Ab(\mathcal{E})$ has enough injectives. The global sections functor $\Gamma : \mathcal{E} \rightarrow \mathcal{S}$ (Section 2) sends abelian group objects to abelian groups, so induces a functor (again denoted) $\Gamma : Ab(\mathcal{E}) \rightarrow Ab$, which is left exact and preserves injectives. For any abelian group object A in $\mathcal{E}$, the cohomology groups $H^n(\mathcal{E}, A)$ are defined as the right derived functors of Γ, i.e.

$$H^n(\mathcal{E}, A) = R^n\Gamma(A) \quad (n \geq 0). \qquad (1)$$

The construction of these groups $H^n(\mathcal{E}, A)$ is functorial, contravariant in $\mathcal{E}$ and covariant in A, as usual. For any object $B \in \mathcal{E}$, one also considers the right derived functors of the functor $\mathrm{Hom}_{\mathcal{E}}(B, -)$, which sends an abelian group A to the group of arrows $B \to A$ in $\mathcal{E}$ ("sections of A over B"). These groups are denoted $H^n(\mathcal{E}, B; A)$. For such an object $B \in \mathcal{E}$, the functor $B^* : \mathcal{E} \to \mathcal{E}/B$ (sending E to $E \times B \to B$, see Section 2) induces a functor $B^* : Ab(\mathcal{E}) \to Ab(\mathcal{E}/B)$ which is exact and preserves injectives. Thus $H^n(\mathcal{E}, B; A) \cong H^n(\mathcal{E}/B, B^*(A))$, and one also denotes the latter group simply by $H^n(\mathcal{E}/B, A)$.

For a topological space X and an abelian sheaf A on X, the topos cohomology groups $H^n(Sh(X), A)$ are the usual sheaf cohomology groups (cf. Godement (1958), Iversen(1986)). For a small category $\mathbf{C}$, the topos cohomology groups $H^n(\mathcal{B}\mathbf{C}, A)$ are (isomorphic to) to cohomology groups of the category $\mathbf{C}$; see Proposition II.6.1, below.

For any topos morphism $f : \mathcal{F} \to \mathcal{E}$, the direct image functor f_* defines a left exact functor $f_* : Ab(\mathcal{F}) \to Ab(\mathcal{E})$, which preserves injectives, and has the property that it respects the global sections functors, in the sense that there is a natural isomorphism $\Gamma(f_*(A)) \cong \Gamma(A)$, for any $A \in Ab(\mathcal{F})$. The Grothendieck spectral sequence (Grothendieck(1957)) for the composite $\Gamma \circ f_*$ is known as the *Leray spectral sequence* for f, and takes the from

$$E_2^{p,q} = H^p(\mathcal{E}, R^q f_*(A)) \;\Rightarrow\; H^{p+q}(\mathcal{F}, A)\,. \tag{2}$$

There is another fundamental spectral sequence, associated to any suitable simplicial object $X. = \{X_p\}_{p \geq 0}$ in a topos $\mathcal{E}$. Such an object $X.$ gives rise to an augmented chain complex in $Ab(\mathcal{E})$,

$$0 \leftarrow \mathbf{Z} \leftarrow \mathbf{Z} \cdot X_0 \stackrel{\partial}{\leftarrow} \mathbf{Z} \cdot X_1 \leftarrow \cdots \tag{3}$$

Here $\mathbf{Z} \cdot (-) : \mathcal{E} \to Ab(\mathcal{E})$ denotes the free abelian group functor, sending an object E to the sum $\sum_{n \in \mathbf{Z}} E$ in $\mathcal{E}$; for $E = 1$, we simply write $\mathbf{Z}$ for $\mathbf{Z} \cdot 1 = \Delta(\mathbf{Z})$; the boundary ∂ is defined in the usual way from the face operators of the simplicial object $X.$ by alternating sums. This object $X.$ is said to be *locally acyclic* if the complex (3) is exact. For any such locally acyclic $X.$, there is a spectral sequence

$$E_2^{p,q} = H^p H^q(\mathcal{E}/X., A) \Rightarrow H^{p+q}(\mathcal{E}, A) \tag{4}$$

Here A is any abelian group object in $\mathcal{E}$, and, as above, we write A for $X_p^*(A)$ in the cohomology $H^q(\mathcal{E}/X_p, A)$.

An important special case of such locally acyclic simplicial objects are the *hypercovers* of $\mathcal{E}$. To define these, recall first from Quillen(1967) that a map $f : Y. \to X.$ between simplicial sets is a *trivial fibration* if any commutative square of the form

$$\begin{array}{ccc} \dot{\Delta}[n] & \longrightarrow & Y. \\ \downarrow & \nearrow & \downarrow \\ \Delta[n] & \longrightarrow & X. \end{array}$$

has a diagonal filling (as indicated by the dotted arrow). Here $\Delta[n]$ is the standard n-simplex and $\dot{\Delta}[n]$ its boundary. Thus f is a trivial fibration precisely when the map

$$X_n = \mathrm{Hom}(\Delta[n], X.) \to \mathrm{Hom}(\dot{\Delta}[n], X.) \times_{\mathrm{Hom}(\dot{\Delta}[n], Y.)} \mathrm{Hom}(\Delta[n], Y.) \tag{5}$$

is a surjective map between sets. If $Y. = 1$, this is the familiar requirement that $X.$ is a *contractible Kan complex.* Call a map $f : Y. \to X.$ between simplicial objects in a topos $\mathcal{E}$ a *local trivial fibration* if the similar map (5) is an epimorphism in $\mathcal{E}$. A *hypercover* of $\mathcal{E}$ is by definition a simplicial object $X.$ in $\mathcal{E}$ so that the map $X. \to 1$ is such a local trivial fibration. (Thus $X.$ is "locally" (or "internally") a contractible Kan complex in $\mathcal{E}$, in some sense.) Every hypercover is locally acyclic, and gives rise to a spectral sequence (4). Denote by $HC(\mathcal{E})$ the category of hypercovers and homotopy classes of maps. One can then form a "Verdier cohomology" direct limit over all hypercovers (a generalized Čech cohomology):

$$\begin{aligned} \check{H}^p_{Verdier}(\mathcal{E}, A) &= \varinjlim_{X. \in HC(\mathcal{E})} H^p \,\mathrm{Hom}_{\mathcal{E}}(X., A) \\ &= \varinjlim_{X.} H^p H^0(\mathcal{E}/X., A)\ . \end{aligned}$$

The direct limit of the spectral sequences (4) collapses, and gives an isomorphism

$$\check{H}^p_{Verdier}(\mathcal{E}, A) \cong H^p(\mathcal{E}, A)\ . \tag{6}$$

The hypercovers of $\mathcal{E}$ are also used to define the ("étale") homotopy groups $\pi_n(\mathcal{E}, p)$ of the topos $\mathcal{E}$ with a chosen base-point p, i.e. a topos morphism $p : \mathcal{S} \to \mathcal{E}$ from the topos $\mathcal{S}$ of sets. Before we give the general definition, we discuss the special case $n = 1$ of the *fundamental group.* The *profinite* fundamental group is discussed in SGA1. The more general case requires the topos to be *locally connected.* To define this notion, first call a non-zero object E of $\mathcal{E}$ *connected* if E cannot be decomposed as a sum $E = E_1 + E_2$, except in the trivial ways where $E_1 = 0$ or $E_2 = 0$. The topos $\mathcal{E}$ is called locally connected if every object E in $\mathcal{E}$ can be decomposed as a sum of connected objects, say $E = \sum_{i \in I} E_i$. This decomposition is essentially unique, and its index set I is the set of connected components of E, denoted $\pi_0(E)$. This construction defines, for any locally connected topos $\mathcal{E}$, a functor

$$\pi_0 : \mathcal{E} \to \mathcal{S}\ ,$$

which is left adjoint to the constant sheaf functor $\Delta : \mathcal{S} \to \mathcal{E}$. For the terminal object 1 of $\mathcal{E}$, one also writes $\pi_0(\mathcal{E})$ for $\pi_0(1)$, and calls this set the set of connected components of $\mathcal{E}$: in particular, $\mathcal{E}$ is a *connected* topos iff $\pi_0(\mathcal{E})$ is a one-point set.

Next, an object E of $\mathcal{E}$ is called *locally constant* if there exists a set S, an epi $U \twoheadrightarrow 1$ in $\mathcal{E}$, and an isomorphism $E \times U \cong \sum_{s \in S} U$ over U. For example, in the case where $\mathcal{E}$ is the topos of sheaves on a space X, an object (étale map) $E \to X$ is locally constant precisely when it is a covering projection. Thus, we also refer to locally constant objects of $\mathcal{E}$ as *covering spaces* of $\mathcal{E}$. For a locally connected topos $\mathcal{E}$, consider the full subcategory $SLC(\mathcal{E})$ of $\mathcal{E}$, consisting of sums of locally constant objects. If $p : \mathcal{S} \to \mathcal{E}$ is a point of $\mathcal{E}$, its inverse image functor restricts to a functor $p^* : SLC(\mathcal{E}) \to \mathcal{S}$. An infinite version of Grothendieck's Galois theory (Artin-Mazur(1969), Moerdijk(1989)) gives an essentially unique progroup G such that there is an equivalence of categories between $SLC(\mathcal{E})$ and the category $\mathcal{B}G$ of sets equipped with an action by G. The point p is needed for the construction of G, and the equivalence identifies $p^* : SLC(\mathcal{E}) \to \mathcal{E}$ with the canonical functor "forget the action": $\mathcal{B}G \to \mathcal{S}$. One denotes G by $\pi_1(\mathcal{E}, p)$, and refers to it as *the fundamental group* of $\mathcal{E}$.

This "enlarged" (when compared to the profinite one) fundamental group has many of the familiar properties of the fundamental group of a topological space. For example, for any abelian group A there is a canonical isomorphism

$$H^1(\mathcal{E}, \Delta(A)) \;\cong\; \mathrm{Hom}(\pi_1(\mathcal{E}, p),\, A)\,,$$

analogous to the Hurewicz theorem for topological spaces which states that the first homology group is the abelianization of the fundamental group.

One can also define higher homotopy groups of a locally connected topos $\mathcal{E}$ with a base-point p. These higher homotopy groups are again progroups, called the étale homotopy groups of $(\mathcal{E}, p)$ and denoted $\pi_n(\mathcal{E}, p)$ (or $\pi_n^{et}(\mathcal{E}, p)$). For $n = 1$, this agrees with the fundamental group just described. The construction of these higher homotopy groups can be outlined as follows. For any hypercover $X.$ of $\mathcal{E}$, the connected components form a simplicial set $\pi_0(X.)$. A base-point of such a hypercover (over the point p of $\mathcal{E}$) is by definition a vertex x_0 of the simplicial set $p^*(X.)$. This vertex x_0 yields a corresponding vertex $\tilde{x}_0$ of $\pi_0(X.)$ – its image under the map $p^*(X.) \to p^*\Delta\pi_0(X.) \cong \pi_0(X.)$ induced by the unit of the adjunction between Δ and π_0. The étale homotopy groups are defined as the progroups ("formal inverse limits")

$$\pi_n(\mathcal{E}, p) \;=\; \underset{\longleftarrow\,(X,x_0)}{\lim}\; \pi_n(\pi_0(X.),\, \tilde{x}_0)\,,$$

indexed by all the pointed hypercovers and homotopy classes of maps between them (or rather, some "small" cofinal subsystem of these).

A topos morphism $f : \mathcal{F} \to \mathcal{E}$ induces for each point $q : \mathcal{S} \to \mathcal{F}$ of $\mathcal{F}$ homomorphisms

$$\pi_n(f) : \pi_n(\mathcal{F}, q) \to \pi_n(\mathcal{E}, fq)\,.$$

As for topological spaces, these depend only on the homotopy class of f. More explicitly, let I denote the unit interval, and also (by the conventions of Section 2) its topos of sheaves. Two morphisms $f_0, f_1 : \mathcal{F} \to \mathcal{E}$ are said to be homotopic if there is a topos morphism $H : I \times \mathcal{F} \to \mathcal{E}$ such that $H \circ i_k \cong f_k$ for $k = 0, 1$ (where $i_0, i_1 : 1 \to I$ are the inclusions of the base-points). The homotopy H is said to be relative to the base-points q of $\mathcal{F}$ and p of $\mathcal{E}$ if the square

$$\begin{array}{ccc} I \times \mathcal{S} & \xrightarrow{I \times q} & I \times \mathcal{F} \\ \downarrow & & \downarrow{\scriptstyle H} \\ \mathcal{S} & \xrightarrow{p} & \mathcal{E} \end{array}$$

commutes up to isomorphism. If so, then $\pi_n(f_0) = \pi_n(f_1)$ as maps between progroups.

Similarly, if f_0 and f_1 are homotopic maps and A is a locally constant abelian group in A, then $f_0^*(A) \cong f_1^*(A)$, and (modulo this isomorphism) $f_0^* = f_1^* : H^n(\mathcal{E}, A) \to H^n(\mathcal{F}, f_i^* A)$.

In particular, this applies to an arrow $\alpha : f_0 \to f_1$ between two topos morphisms, i.e. a natural transformation $\alpha : f_0^* \to f_1^*$, because such an α can be interpreted as a "natural" homotopy. Indeed, let Σ be the Sierpinski space, i.e., the two point space with an open point 1 and a closed point 0. Its topos $Sh(\Sigma)$ of sheaves is simply the arrow category of the category of sets. Similarly, the product $Sh(\Sigma) \times \mathcal{F}$ is the topos with as objects arrows $F_0 \to F_1$ in $\mathcal{F}$ and as arrows the commutative squares. Thus α is a "Sierpinski homotopy" $Sh(\Sigma) \times \mathcal{F} \to \mathcal{E}$. By composition with a continuous surjection $p : I \to \Sigma$ which preserves the endpoints, one obtains a homotopy between f_0 and f_1. Thus, if topos morphisms f_0 and f_1 are related by a natural transformation, they operate identically on cohomology with locally constant coefficients, and on étale homotopy (if the natural transformation respects the base-points).

As usual, we will denote the collection of homotopy classes of topos morphisms from $\mathcal{F}$ to $\mathcal{E}$ by

$$[\mathcal{F}, \mathcal{E}].$$

For a topological space X with a base-point x_0, one thus has the usual homotopy groups $\pi_n(X, x_0)$ and the étale homotopy groups $\pi_n^{et}(Sh(X), x_0)$ of the topos of sheaves (with point $x_0 : \mathcal{S} \to Sh(X)$ corresponding to $x_0 \in X$). If the space X has a basis of contractible open sets, then these progroups are actually ordinary groups, and there is a natural isomorphism (Artin-Mazur(1969), Section 12)

$$\pi_n^{et}(Sh(X), x_0) \;\cong\; \pi_n(X, x_0)\,. \tag{7}$$

A topos morphism $f : \mathcal{F} \to \mathcal{E}$ between locally connected topoi is said to be a *weak homotopy equivalence* if f induces an isomorphism $\pi_0(\mathcal{F}) \to \pi_0(\mathcal{E})$ between the sets of connected components, and, for any base-point q of $\mathcal{F}$, isomorphisms $\pi_n(\mathcal{F}, q) \xrightarrow{\sim} \pi_n(\mathcal{E}, fq)$ (for $n \geq 1$). By the "toposophic Whitehead theorem" (Artin-Mazur(1969), Section 4) f is a weak homotopy equivalence iff f induces isomorphisms for π_0 and π_1, and for each locally constant abelian group A in $\mathcal{E}$ an isomorphism $f^* : H^n(\mathcal{E}, A) \xrightarrow{\sim} H^n(\mathcal{F}, f^*A)$ (for $n \geq 0$).

In many examples, f will have the property that $f^* : \mathcal{F} \to \mathcal{E}$ is full and faithful. Such an f is called a *connected morphism.* Any such connected morphism induces an isomorphism $\pi_0(\mathcal{F}) \xrightarrow{\sim} \pi_0(\mathcal{E})$ of connected components, and a surjection of fundamental groups.

Chapter II

Classifying Topoi

§1 Group actions

Let G be a (discrete) group, and let $\mathcal{B}G$ be the topos of right G-sets. This is a special case of the presheaf topos $\mathcal{B}\mathbb{C}$ introduced in Chapter I, Section 2, when G is viewed as a category with one object. As a simple and motivating example of a classifying topos, we will describe in this section how the topos $\mathcal{B}G$ "classifies" principal G-bundles.

Recall that for a topological space X, a principal G-bundle on X is a surjective sheaf $p : E \to X$, equipped with a continuous fiberwise left G-action $\alpha : G \times E \to E$ (denoted $\alpha(g, e) = g \cdot e$) which is free and transitive on each fiber. Thus the map

$$(\alpha, \pi_2) : G \times E \to E \times_X E$$

is a homeomorphism (of étale spaces over X). It follows that the map $p : E \to X$ must be a covering projection.

A map between two such principal bundles $E = (E, p, \alpha)$ and $E' = (E', p', \alpha')$ is a map $\varphi : E \to E'$ of sheaves on X which preserves the action. Any such map φ must be an isomorphism. This defines a category $\mathrm{Prin}(X, G)$ of principal G-bundles over X.

1.1. Proposition. *There is a natural equivalence of categories*

$$\mathrm{Hom}(X, \mathcal{B}G) \tilde{\to} \mathrm{Prin}(X, G) .$$

Note that, according to the conventions of Chapter I, Section 2, the X on the left of this equivalence stands for the topos $Sh(X)$ of all sheaves.

We prove the proposition, and make some further comments on naturality after the proof.

Proof. Let $f : X \to \mathcal{B}G$ be a morphism of topoi. Consider the right G-set $\tilde{G}$, given by G acting on itself by multiplication from the right. This $\tilde{G}$ is an object of $\mathcal{B}G$. Let

$$E = f^*(\tilde{G}) .$$

This sheaf E is surjective; indeed, $\tilde{G} \times \tilde{G} \rightrightarrows \tilde{G} \to 1$ is a coequalizer in $\mathcal{B}G$, and f^* preserves products and coequalizers, so $E \times_X E \rightrightarrows E \to X$ is a coequalizer of spaces.

For each $g \in G$, the left multiplication $\lambda_g(x) = g \cdot x$ defines a map $\lambda_g : \tilde{G} \to \tilde{G}$ in the category $\mathcal{B}G$. Thus one obtains a map $f^*(\lambda_g) : E \to E$ of sheaves. For a point $y \in E$, write $g \cdot y = f^*(\lambda_g)(y)$. This defines an action α of G on E. To see that it is free and transitive, note that the map

$$\bar{\lambda} : \sum_{g \in G} \tilde{G} \to \tilde{G} \times \tilde{G} \quad \bar{\lambda}_g(x) = (g \cdot x, x)$$

is an isomorphism in $\mathcal{B}G$. Since f^* preserves sums, products and isomorphisms, it sends this map $\bar{\lambda}$ into an isomorphism

$$\bar{\alpha} : \sum_{g \in G} E \xrightarrow{\sim} E \times_X E \quad \bar{\alpha}_g(y) = (g \cdot y, y) .$$

This means precisely that the action by G on E is principal.

Conversely, suppose $p : E \to X$ is a principal G-bundle over X. If S is any object from $\mathcal{B}G$ (i.e., a right G-set), consider the "tensor-product"

$$S \otimes_G E$$

(also often denoted $S \times_G E$), obtained from $S \times E$ by the identifications $(s \cdot g, e) \sim (s, g \cdot e)$. We denote equivalence classes by $s \otimes e$. The natural map $p_S : S \times_G E \to X$, $p_S(s \otimes e) = p(e)$, is a well-defined local homeomorphism. Thus $S \otimes_G E$ is a sheaf on X. The construction is evidently functorial in S, so this defines a functor

$$- \otimes_G E \; : \; \mathcal{B}G \to Sh(X) .$$

To see that this functor preserves colimits and finite limits, it suffices to check this for the stalk at each point $x \in X$. But for a G-set S,

$$(S \otimes_G E)_x \cong S \otimes_G E_x \cong S ,$$

where the latter isomorphism is natural in S but depends on the point x in a non-canonical way: choose $y \in E_x$ – then $s \mapsto s \otimes y$ is an isomorphism $S \to S \otimes_G E_x$, precisely because the G-action on the stalk E_x is free and transitive. In any case, since $(S \otimes_G E)_x \cong S$ for each point x, it is clear that $- \otimes_G E$ preserves colimits and finite limits. Thus, as explained in Chapter I, this functor is the inverse image part of a topos morphism $X \to \mathcal{B}G$, uniquely determined up to isomorphism.

Finally, it is straightforward to check that these constructions, from a principal bundle out of a topos morphism and conversely, are mutually inverse up to natural isomorphism. In outline, for any right G-set S there is a canonical isomorphism

$$S \otimes_G \tilde{G} \cong S$$

of right G-sets. For a given morphism $f : X \to \mathcal{B}G$ this gives the required natural isomorphism $- \otimes_G E \cong f^*$ for $E = f^*(\tilde{G})$. Conversely, for any principal bundle E, there is a canonical isomorphism of principal bundles $\tilde{G} \otimes_G E \cong E$.

The equivalence in the statement of the proposition is both natural in X and in G. For X, this means that for any continuous map $\varphi : Y \to X$ between spaces, the square

$$\begin{array}{ccc} \mathrm{Hom}(X, \mathcal{B}G) & \xrightarrow{\sim} & \mathrm{Prin}(X, G) \\ {\scriptstyle \varphi^*}\downarrow & & \downarrow{\scriptstyle \varphi^*} \\ \mathrm{Hom}(Y, \mathcal{B}G) & \xrightarrow{\sim} & \mathrm{Prin}(Y, G). \end{array}$$

commutes, where φ^* on the left denotes "compose with φ", while φ^* on the right is "pullback". For G, it means that for any homomorphism of groups $\psi : G \to H$, there is a commutative square

$$\begin{array}{ccc} \mathrm{Hom}(X, \mathcal{B}G) & \xrightarrow{\sim} & \mathrm{Prin}(X, G) \\ {\scriptstyle \psi_*}\downarrow & & \downarrow{\scriptstyle \psi_!} \\ \mathrm{Hom}(X, \mathcal{B}H) & \xrightarrow{\sim} & \mathrm{Prin}(X, H). \end{array}$$

Here ψ_* on the left is given by composition with the morphism $\mathcal{B}G \to \mathcal{B}H$ induced by ψ (see Section I.2). On the right, $\psi_!$ is defined for any principal G-bundle E by

$$\psi_!(E) = H \otimes_G E \ ,$$

where H is viewed as a right G-set with action α defined by $\alpha(h, g) = h \cdot \psi(g)$.

Recall from Chapter I, Section 4, that $[X, \mathcal{B}G]$ denotes the collection of homotopy classes of topos morphisms.

1.2. Corollary. *There is a natural bijection between $[X, \mathcal{B}G]$ and the collection of isomorphism classes of principal bundles.*

Proof. This follows from the equivalence of Proposition 1.1. For, on the one hand, if f and $g : X \to \mathcal{B}G$ are homotopic maps, then the corresponding principal G-bundles $E = f^*(\tilde{G})$ and $F = g^*(\tilde{G})$ are "concordant"; i.e. there is a principal bundle H on $X \times [0,1]$ so that $E \cong H|X \times \{0\}$ and $F \cong H|X \times \{1\}$. Since every principal G-bundle over $[0,1]$ is constant, it follows that E and F are isomorphic. Conversely, on the other hand, if $\alpha : E \to F$ is an (iso-)morphism between principal G-bundles over X, then α corresponds under the equivalence of Proposition 1.1 to a natural transformation between the classifying maps f and $g : X \to \mathcal{B}G$. Thus, as explained in Section I.4, these maps are homotopic.

Although we will not use this more general case, it should be noted that for any topos $\mathcal{E}$, one can define the notion of a principal G-bundle over $\mathcal{E}$, and prove an equivalence $\mathrm{Hom}(\mathcal{E}, \mathcal{B}G) \xrightarrow{\sim} \mathrm{Prin}(\mathcal{E}, G)$, analogous to Proposition 1.1, in exactly the same way.

§2 Diaconescu's theorem

In this section, we will extend Proposition 1.1 from groups to arbitrary (small) categories. Let $\mathbb{C}$ be such a category, with classifying topos $\mathcal{B}\mathbb{C}$ of all presheaves on $\mathbb{C}$, as described in Chapter I, Section 2. For a topological space X, a $\mathbb{C}$ *-bundle* over X is a covariant functor $E : \mathbb{C} \to Sh(X)$. In other words, a $\mathbb{C}$-bundle consists of a sheaf $E(c)$ for each object c in $\mathbb{C}$, and a sheaf map $E(\alpha) : E(c) \to E(d)$ for each arrow $\alpha : c \to d$, denoted

$$E(\alpha)(y) = \alpha \cdot y \qquad (y \in E(c)) ,$$

so that the usual identities $\mathrm{id}_c \cdot y = y$ and $\beta \cdot (\alpha \cdot y) = (\beta\alpha) \cdot y$ are satisfied. Such a $\mathbb{C}$-bundle E is said to be *principal* (or *flat*, or *filtering*) if for each point $x \in X$ the following conditions are satisfied for the stalks $E(c)_x$:

(i) (non-empty) There is at least one object $c \in \mathbb{C}$ for which the stalk $E(c)_x$ is non-empty.

(ii) (transitive) For any two points $y \in E(c)_x$ and $z \in E(d)_x$, there are arrows $\alpha : b \to c$ and $\beta : b \to d$ from some object $b \in \mathbb{C}$, and a point $w \in E(b)_x$, so that $\alpha \cdot w = y$ and $\beta \cdot w = z$.

(iii) (free) For any two parallel arrows $\alpha, \beta : c \rightrightarrows d$ and any $y \in E(c)_x$ for which $\alpha \cdot y = \beta \cdot y$, there exists an arrow $\gamma : b \to c$ and a point $z \in E(b)_x$ so that $\alpha\gamma = \beta\gamma$ and $\gamma \cdot z = y$.

2.1. Examples. (a) A group G can be viewed as a one-object category. In this case the above notion of principal $\mathbb{C}$-bundle agrees with the usual one, discussed in the previous section.

(b) Let M be a monoid, again viewed as a one-object category. Then M is said to have right cancellation if $km = \ell m$ implies $k = \ell$, for any $k, \ell, m \in M$ (in other words, if every arrow in M is epi). In this case, a principal M-bundle over X is a surjective sheaf $E \to X$ with a continuous (fiberwise) left action by M, which is free in the sense that $m \cdot e = n \cdot e$ implies $m = n$ (for any $e \in E$), and transitive as in condition (ii) above. These are exactly the principal M-bundles discussed in Segal (1978), p.378 (except that Segal considers right actions, hence assumes that M has left cancellation).

(c) A partially ordered set P can be viewed as a category, with exactly one arrow $p \to q$ iff $p \leq q$. In this case, a principal P-bundle over X is a family $\{U_p : p \in P\}$ of open subsets in X with the following properties: if $p \leq q$ then $U_p \subseteq U_q$; the U_p together cover X; the U_p are "locally" directed, in the sense that $U_p \cap U_q$ is covered by all U_r with $r \leq p$ and $r \leq q$.

(d) If the category $\mathbb{C}$ has finite limits, then a $\mathbb{C}$-bundle $E : \mathbb{C} \to Sh(X)$ over X is principal iff it sends all finite limits in $\mathbb{C}$ to finite limits in $Sh(X)$. More generally, for any small category $\mathbb{C}$, a principal bundle E must commute with all those finite limits which exist in $\mathbb{C}$. (See Mac Lane-Moerdijk(1992), Chapter VII.)

For two principal $\mathbb{C}$-bundles E and E' over X, a morphism $\varphi : E \to E'$ is by definition a natural transformation; i.e. φ is a family of sheaf maps $\varphi_c : E(c) \to E'(c)$ (for $c \in \mathbb{C}$) so that $\varphi_d(\alpha \cdot y) = \alpha \cdot \varphi_c(y)$ for any $\alpha : c \to d$ in $\mathbb{C}$ and any point $y \in E(c)$. In this way, the principal $\mathbb{C}$-bundles over X form a category, denoted

$$\mathrm{Prin}(X, \mathbb{C}) .$$

2.2. Theorem. *For any small category* $\mathbb{C}$ *and any topological space* X*, there is a natural equivalence of categories*

$$\mathrm{Hom}(X, \mathcal{B}\mathbb{C}) \;\cong\; \mathrm{Prin}(X, \mathbb{C}) .$$

Proof. The proof will follow the same pattern as that of Proposition 1.1. In one direction, let $f : X \to \mathcal{B}\mathbb{C}$ be a morphism of topoi. By composition with the Yoneda embedding $\mathrm{Yon} : \mathbb{C} \to \mathcal{B}\mathbb{C}$, one obtains a functor

$$E = f^* \circ \mathrm{Yon} : \mathbb{C} \to Sh(X) .$$

To see that E is a principal bundle, we verify conditions (i)-(iii). For condition (i), note that $\Sigma_{c\in\mathbb{C}}\ \mathrm{Yon}(c) \to 1$ is an epimorphism to the terminal object 1 in $\mathcal{B}\mathbb{C}$. Since f^* preserves epis and sums, $\Sigma_{c\in\mathbb{C}}\ E(c) \to X$ must be surjective. For condition (ii), observe that for any two objects c and d of $\mathbb{C}$, the evident map

$$\sum\nolimits_{c\leftarrow b\rightarrow d} \mathrm{Yon}(b) \;\to\; \mathrm{Yon}(c) \times \mathrm{Yon}(d)$$

is an epimorphism in $\mathcal{B}\mathbb{C}$. Applying f^* thus yields an epimorphism of sheaves

$$\sum\nolimits_{c\leftarrow b\rightarrow d} E(b) \;\to\; E(c) \times E(d) ,$$

so that condition (ii) is satisfied. Finally, for $\alpha, \beta : c \rightrightarrows d$, condition (iii) follows similarly, by applying f^* to the equalizer diagram

$$\sum_{\substack{\gamma : b \to c \\ \alpha\gamma = \beta\gamma}} \mathrm{Yon}(b) \;\to\; \mathrm{Yon}(c) \;\rightrightarrows\; \mathrm{Yon}(d)$$

in $\mathcal{B}\mathbb{C}$.

Conversely, suppose E is a principal $\mathbb{C}$-bundle over X. For any presheaf S on $\mathbb{C}$, one can define a "tensor product"

$$S \otimes_{\mathbb{C}} E ;$$

This is the quotient of the sum of sheaves $\sum_{c\in\mathbb{C}, s\in S(c)} E(c)$, obtained by the identifications

$$(s \cdot \alpha, e) \;\sim\; (s, \alpha \cdot e)$$

for any $\alpha : c \to d$, $s \in S(d)$ and $e \in E(c)$. Again, we denote the points of $S \otimes_{\mathbb{C}} E$ by $s \otimes e$. This construction defines a functor

$$- \otimes_{\mathbb{C}} E \;:\; \mathcal{B}\mathbb{C} \;\to\; Sh(X) ,$$

which clearly commutes with colimits (since $S \otimes_{\mathbf{C}} E$ is itself constructed as a colimit). The assumption that E is principal will ensure that $- \otimes_{\mathbf{C}} E$ is also left exact. Indeed, to prove this, it suffices to show that for any point $x \in X$, the stalk functor

$$S \mapsto (S \otimes_{\mathbf{C}} E)_x \;\cong\; S \otimes_{\mathbf{C}} E_x$$

is left exact, i.e. commutes with finite limits. To this end, let I_x be the category with as objects pairs (c, y) where $y \in E_x(c)$, and as arrows $\alpha : (c, y) \to (d, z)$ those arrows $\alpha : c \to d$ in $\mathbf{C}$ for which $\alpha \cdot y = z$. Then the conditions that E is principal exactly mean that each category I_x^{op} (the dual of I_x) is a filtering category (Mac Lane(1971), p. 207). Furthermore, a presheaf S on $\mathbf{C}$ gives by composition a functor $\tilde{S}_x : I_x^{op} \to \mathbf{C}^{op} \to (sets)$, and there is a canonical isomorphism

$$S \otimes_{\mathbf{C}} E_x \;\cong\; \varinjlim_{I_x^{op}} \tilde{S}_x \,.$$

Thus $S \mapsto S \otimes_{\mathbf{C}} E_x$ preserves finite limits, since these commute with filtered colimits (Mac Lane (1971), p. 211). This proves that the functor $- \otimes_{\mathbf{C}} E : \mathcal{B}\mathbf{C} \to Sh(X)$ commutes with colimits and finite limits, and hence is the inverse image functor of a topos morphism $X \to \mathcal{B}\mathbf{C}$, uniquely determined up to isomorphism.

To complete the proof of the theorem, it must be verified that these constructions, of a principal bundle from a topos morphism, and of a topos morphism out of a principal bundle, are mutually inverse up to natural isomorphism.

In one direction, start with a principal bundle E, construct a morphism $f : X \to \mathcal{B}\mathbf{C}$ with $f^* = - \otimes_{\mathbf{C}} E$, and define a new principal bundle E' by $E'(c) = f^*(\mathrm{Yon}(c)) = \mathrm{Yon}(c) \otimes_{\mathbf{C}} E$. The evident map $\sigma : E'(c) = \mathrm{Yon}(c) \otimes_{\mathbf{C}} E \to E(c)$, defined by the formula $\sigma([\alpha : b \to c] \otimes z) = \alpha \cdot z$, is clearly an isomorphism (representable functors are "units" for the tensor product).

The other way round, start with a morphism f and construct $E = f^* \circ \mathrm{Yon}$, and then a new morphism with inverse image functor $- \otimes_{\mathbf{C}} E$. For each presheaf S on $\mathbf{C}$, there is a canonical map

$$\tau : S \otimes_{\mathbf{C}} E \to f^*(S) \;,\quad \tau(s \otimes e) = f^*(\hat{s})(\dot{z})$$

(here $s \in S(c)$, with corresponding map $\hat{s} : \mathrm{Yon}(c) \to S$, and $z \in E(c)$). If S is itself representable, say $S = \mathrm{Yon}(d)$, the τ is the standard isomorphism $\mathrm{Yon}(d) \otimes_{\mathbf{C}} E \cong E(d)$. Since every presheaf S is a colimit of representables while τ is natural in S, it follows that τ is an isomorphism for each S.

This proves the theorem.

2.3. Remark. Just as in Proposition 1.1, the equivalence of Theorem 2.2 is again natural in X and $\mathbf{C}$. For a map $f : Y \to X$ of spaces and a functor $\varphi : \mathbf{C} \to \mathbf{D}$, this is expressed by the following two squares which commute (up to isomorphism):

$$\begin{array}{ccc}
\mathrm{Hom}(Y, \mathcal{B}\mathbf{C}) & \xrightarrow{\sim} & \mathrm{Prin}(Y, \mathbf{C}) \\
\uparrow & & \uparrow f^* \\
\mathrm{Hom}(X, \mathcal{B}\mathbf{C}) & \xrightarrow{\sim} & \mathrm{Prin}(X, \mathbf{C}) \\
\downarrow & & \downarrow \varphi_! \\
\mathrm{Hom}(X, \mathcal{B}\mathbf{D}) & \xrightarrow{\sim} & \mathrm{Prin}(X, \mathbf{D})
\end{array}$$

Here the vertical maps on the left are defined by composition with $f : Y \to X$ and $\varphi : \mathcal{B}\mathbf{C} \to \mathcal{B}\mathbf{D}$, respectively. On the right, f^* is the operation of pulling back principal bundles, whereas $\varphi_!$ is described as follows: for a principal bundle E on X, and an object $d \in \mathbf{D}$,

$$\varphi_!(E)(d) \;=\; \varphi^*(\mathrm{Yon}(d)) \otimes_{\mathbf{C}} E \;,$$

Where $\varphi^*(\mathrm{Yon}(d)) = \mathrm{Yon}(d) \circ \varphi \;=\; \mathbf{D}(\varphi(-), d) \;:\; \mathbf{C}^{op} \to (sets)$ is the inverse image of the representable presheaf $\mathrm{Yon}(d)$ on $\mathbf{D}$ (cf. Chapter I, Section 2).

Next, call two principal $\mathbf{C}$-bundles E_0 and E_1 on a space X *concordant* if there exists a principal bundle E on $X \times [0,1]$ so that $E_0 \cong i_0^*(E)$ and $E_1 \cong i_1^*(E)$ (where $i_0, i_1 : X \hookrightarrow X \times [0,1]$ are the evident inclusions). This defines an equivalence relation on principal bundles. Write

$$k_{\mathbf{C}}(X)$$

for the collection of equivalence classes ("concordance classes") of principal bundles. Note that if $\varphi : E_0 \to E_1$ is a map between principal $\mathbf{C}$-bundles on X, then E_0 and E_1 are concordant: one can construct a concordance E on $X \times [0,1]$, with $E|X \times (0,1] \cong \pi^*(E_1)$ for the projection $\pi : X \times [0,1] \to X$, and $E|X \times \{0\} \cong \pi^*(E_0)$, by glueing a point $z \in E_0(c)_x$ to the section $\varphi(z)$ on $\{x\} \times (0,1]$.

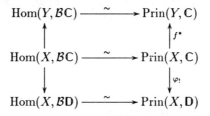

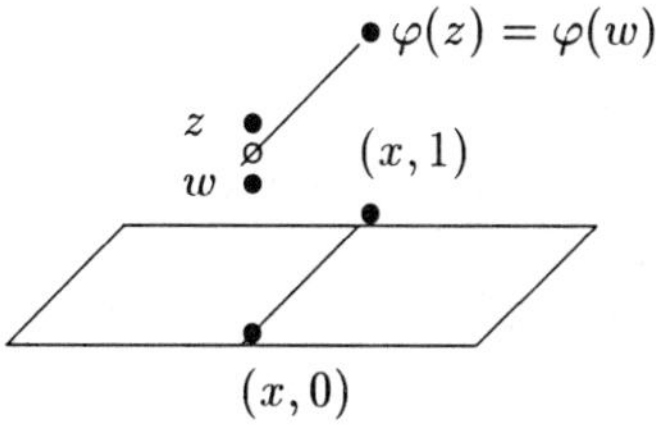

(Under the equivalence of Proposition 2.2, this is really the construction of a homotopy from a natural homotopy in Chapter I, Section 4.)

2.4. Corollary. *For any space X and any small category $\mathbf{C}$, there is a natural isomorphism*

$$[X, \mathcal{B}\mathbf{C}] \cong k_{\mathbf{C}}(X) \,.$$

Proof. Immediate from Prop. 2.2.

2.5. Remark. For any topos $\mathcal{E}$, there is an equivalence between morphisms $\mathcal{E} \to \mathcal{B}\mathbb{C}$ and principal $\mathbb{C}$-bundles over $\mathcal{E}$, similar to Proposition 2.2. (This is discussed in detail in Mac Lane-Moerdijk (1992), Chapter VII.) This more general version is often referred to as "Diaconescu's theorem" (Diaconescu (1975)).

§3 The classifying topos of a topological category

Let $\mathbb{C}$ be a topological category (this notion is discussed in detail, e.g., in Segal(1968) and Bott(1972)). Thus $\mathbb{C}$ is given by a space $\mathbb{C}_0$ of objects and a space $\mathbb{C}_1$ of arrows, and the structure maps for a category are all continuous. We will often denote these maps by $s, t : \mathbb{C}_1 \rightrightarrows \mathbb{C}_0$ for source and target, $m : \mathbb{C}_1 \times_{\mathbb{C}_0} \mathbb{C}_1 \to \mathbb{C}_1$ for composition and $u : \mathbb{C}_0 \to \mathbb{C}_1$ for units; $m(f,g)$ is also denoted $f \circ g$, while we often write 1_x or id_x for $u(x)$.

A $\mathbb{C}$-*sheaf* is a sheaf (étale space) $p : S \to \mathbb{C}_0$ equipped with a continuous right action $\alpha : S \times_{\mathbb{C}_0} \mathbb{C}_1 \to S$, denoted $\alpha(x,f) = x \cdot f$. Thus $x \cdot f$ is defined whenever $p(x) = t(f)$, and satisfied the usual identities for an action:

$$(x \cdot f) \cdot g = x \cdot (f \circ g) \quad , \quad x \cdot 1_{p(x)} = x \quad , \quad p(x \cdot f) = s(f).$$

A map between $\mathbb{C}$-sheaves is defined to be a map of sheaves over $\mathbb{C}_0$ which respects the action. This defines a category of $\mathbb{C}$-sheaves, denoted

$$\mathcal{B}\mathbb{C}\,. \tag{1}$$

3.1. Examples. (a) A small category $\mathbb{C}$ as considered in Section I.2 can be viewed as a topological category with the discrete topology. In this case, a $\mathbb{C}$-sheaf is the same thing as a presheaf on $\mathbb{C}$, and the notation $\mathcal{B}\mathbb{C}$ in (1) is consistent with the one introduced in Chapter I, Section 2.

(b) A topological space X can be viewed as a topological category $\mathbf{X}$, in which all arrows are identities. (So $\mathbf{X}_0 = X = \mathbf{X}_1$, etc.) An $\mathbf{X}$-sheaf is a sheaf on X (Chapter I, Section 2), and $\mathcal{B}\mathbf{X} = Sh(X)$.

(c) Let G be a topological group acting from the right on a space X. Let X_G be the associated translation category: it has X as space of objects, and $X \times G$ as space of arrows, where (x,g) is an arrow $x \cdot g \to x$. An X_G-sheaf is a sheaf $p : S \to X$ on X, with an action by G on S so that p is G-equivariant. Thus $\mathcal{B}(X_G)$ is the category of *G-equivariant sheaves* on X.

(d) As a special case of (c), assume X is a point and G is connected. Thus $G = X_G$ is a one-object topological category. Since the action of a connected group on a discrete set must be trivial, $\mathcal{B}G$ collapses to the category of sets.

(e) A map $f : Y \to X$ between topological spaces gives rise to a topological category $K(F)$ with Y as space of objects and $Y \times_X Y$ as space of arrows. There is a unique arrow (y, y') from y' to y in $K(f)$ iff $f(y) = f(y')$. An action of this category on a sheaf $S \to Y$ is also called *descent data* on S. For example, if $E \to X$ is any sheaf on X, the pullback sheaf $f^*(E) = E \times_X Y \to Y$ has an evident such action, defined by $(e, y) \cdot (y, y') = (e, y')$. This defines a functor $Sh(X) \to \mathcal{B}(K(f))$. When this functor is an equivalence of categories, the map f is said to be an effective descent map (for sheaves). This is the case, for example, when f is an open or proper surjection.

3.2. Proposition. *The category $\mathcal{B}\mathbf{C}$ of $\mathbf{C}$-sheaves is a topos, called the* classifying topos *of the topological category $\mathbf{C}$.*

Proof. We will prove this proposition under the assumption (true in all cases to be considered later) that the source map $s : \mathbf{C}_1 \to \mathbf{C}_0$ is an open map; see also Remark 3.3 below.

For two $\mathbf{C}$-sheaves $S \to \mathbf{C}_0$ and $T \to \mathbf{C}_0$, their product $S \times_{\mathbf{C}_0} T$ in $Sh(\mathbf{C}_0)$ has a unique $\mathbf{C}$-action making the projections $S \times_{\mathbf{C}_0} T \to S$ and $S \times_{\mathbf{C}_0} T \to T$ $\mathbf{C}$-equivariant. With this action, $S \times_{\mathbf{C}_0} T$ is the product of S and T in the category $\mathcal{B}\mathbf{C}$. In other words, products in $\mathcal{B}\mathbf{C}$ can be constructed as products in $Sh(\mathbf{C}_0)$. Exactly the same applies to other finite limits, sums, and coequalizers of equivalence relations, occurring in the Giraud axioms (G1-3) for a topos. Thus $\mathcal{B}\mathbf{C}$ inherits all the relevant exactness properties, expressed in these axioms, from $Sh(\mathbf{C}_0)$. It remains to be shown that $\mathcal{B}\mathbf{C}$ has a set of generators. For any open subset $U \subseteq \mathbf{C}_0$, the space $t^{-1}(U) \subseteq \mathbf{C}_1$ is equipped with a natural action by $\mathbf{C}$, given by the source map $s : t^{-1}(U) \to \mathbf{C}_0$ and the composition

$$t^{-1}(U) \times_{\mathbf{C}_0} \mathbf{C}_1 \to t^{-1}(U)\ .$$

Thus, $s : t^{-1}(U) \to \mathbf{C}_0$ would be a $\mathbf{C}$-sheaf if s were an étale map. Let $\mathcal{G}$ be the collection of all $\mathbf{C}$-sheaves G for which there exists a surjective $\mathbf{C}$-equivariant map $t^{-1}(U) \twoheadrightarrow G$. This collection is small, since there is only a set of such open U, only a set of equivalence relations to put on $t^{-1}(U)$, and only a set of possible topologies to put on the quotient $G = t^{-1}(U)/R$. To see that $\mathcal{G}$ generates $\mathcal{B}\mathbf{C}$, take an arbitrary $\mathbf{C}$-sheaf $p : S \to G_0$, and a point $y \in S$. Let U be an open subset of $\mathbf{C}_0$ on which there exists a section $\sigma : U \to S$ through y, say $\sigma(x) = y$. Let $\varphi : t^{-1}(U) \to S$ be the map $\varphi(g) = \sigma(tg) \cdot g$, and let $G \subseteq S$ be the image of φ. Thus G is closed under the action by $\mathbf{C}$ on S. Furthermore, since $p : S \to \mathbf{C}_0$ is étale while $s : t^{-1}(U) \to \mathbf{C}_0$ is (assumed) open, the identity $p \circ \varphi = s$ implies that φ is open. Thus G is an open subset of S, hence itself a $\mathbf{C}$-sheaf, which obviously belongs to the collection $\mathcal{G}$. This shows that any $\mathbf{C}$-sheaf S is the union of $\mathbf{C}$-sheaves in $\mathcal{G}$, so that $\mathcal{G}$ generates $\mathcal{B}\mathbf{C}$.

This proves the proposition.

3.3. Remark. The category $\mathcal{B}\mathbf{C}$ can be constructed, as a (bicategorical) colimit, from the topoi $Sh(\mathbf{C}_0), Sh(\mathbf{C}_1), Sh(\mathbf{C}_1 \times_{\mathbf{C}_0} \mathbf{C}_1), \cdots$. Thus, the fact that $\mathcal{B}\mathbf{C}$ is a topos is a special case of the existence of such colimits of topoi, first proved in Moerdijk(1988), Section 2. (See also Makkai-Paré (1989), p. 108.)

The construction of the classifying topos $\mathcal{B}\mathbf{C}$ is functorial in $\mathbf{C}$. More precisely, a continuous functor $\varphi : \mathbf{C} \to \mathbf{D}$ between topological categories induces by pullback along $\varphi : \mathbf{C}_0 \to \mathbf{D}_0$ an evident functor

$$\varphi^* : \mathcal{B}\mathbf{D} \to \mathcal{B}\mathbf{C}\,.$$

Since colimits and finite limits in $\mathcal{B}\mathbf{C}$ and $\mathcal{B}\mathbf{D}$ are computed as colimits and finite limits of underlying sheaves (as in the proof of Proposition 3.2), φ^* commutes with these. Thus φ^* is the inverse image functor of a morphism of topoi $\varphi : \mathcal{B}\mathbf{C} \to \mathcal{B}\mathbf{D}$. Furthermore, exactly as for discrete categories discussed in Chapter I, Section 2, a continuous natural transformation $\tau : \varphi \to \psi$ between two such functors $\varphi, \psi : \mathbf{C} \rightrightarrows \mathbf{D}$ induces a map $\tilde{\tau}$ between topos morphisms $\varphi, \psi : \mathcal{B}\mathbf{C} \rightrightarrows \mathcal{B}\mathbf{D}$. In particular, if $\varphi : \mathbf{C} \to \mathbf{D}$ is an equivalence of topological categories, so that there are $\chi : \mathbf{D} \to \mathbf{C}$ and continuous natural isomorphisms $\varphi\chi \cong id_{\mathbf{D}}$ and $\chi\varphi \cong id_{\mathbf{C}}$, then the induced map $\varphi : \mathcal{B}\mathbf{C} \to \mathcal{B}\mathbf{D}$ is an equivalence of topoi.

More generally, a continuous functor $\varphi : \mathbf{C} \to \mathbf{D}$ is said to be *fully faithful* if the square

$$\begin{array}{ccc} \mathbf{C}_1 & \xrightarrow{\varphi} & \mathbf{D}_1 \\ {\scriptstyle (s,t)}\downarrow & & \downarrow{\scriptstyle (s,t)} \\ \mathbf{C}_0 \times \mathbf{C}_0 & \xrightarrow{\varphi \times \varphi} & \mathbf{D}_0 \times \mathbf{D}_0 \end{array}$$

is a fibered product. It is said to be *essentially surjective* if, for the subspace $Iso(\mathbf{D}) \subseteq \mathbf{D}_1$ of invertible arrows, the map from the pullback along $t : Iso(\mathbf{D}) \to \mathbf{D}_0$,

$$s \circ \pi_2 \ : \ \mathbf{C}_0 \times_{\mathbf{D}_0} Iso(\mathbf{D}) \to \mathbf{D}_0$$

is an open (or proper) surjection. This condition expresses in a strong sense that for any object y of $\mathbf{D}$ there is an isomorphism $y \to \varphi(x)$ for some object x in $\mathbf{C}$. The functor φ is said to be a (categorical) *weak equivalence* if φ is both fully faithful and essentially surjective.

3.4. Proposition. *For any categorical weak equivalence $\varphi : \mathbf{C} \to \mathbf{D}$ for the pullback above, the induced map $\mathcal{B}\mathbf{C} \to \mathcal{B}\mathbf{D}$ is an equivalence of topoi.*

Proof. Write $\mathbf{P}_0 = \mathbf{C}_0 \times_{\mathbf{D}_0} Iso(\mathbf{D})$ for the pullback above, with maps $s\pi_2 : \mathbf{P}_0 \to \mathbf{D}_0$ and $\pi_1 : \mathbf{P}_0 \to \mathbf{C}_0$. One can make $\mathbf{P}_0$ into the space of objects of a topological category $\mathbf{P}$, by defining $\mathbf{P}_1$ as the fibered product

$$\begin{array}{ccc} \mathbf{P}_1 & \longrightarrow & \mathbf{D}_1 \\ \downarrow & & \downarrow \\ \mathbf{P}_0 \times \mathbf{P}_0 & \xrightarrow{s\pi_2 \times s\pi_2} & \mathbf{D}_0 \times \mathbf{D}_0 . \end{array}$$

Thus, the objects of $\mathbf{P}$ are of the form $(x, \alpha : y \tilde{\to} \varphi\,(x))$ where x is an object in $\mathbf{C}$ and α an isomorphism in $\mathbf{D}$, and the arrows from $(x, \alpha : y \tilde{\to} \varphi\,(x))$ to $(x', \alpha' : y' \tilde{\to} \varphi\,(x'))$ are simply arrows $\beta : y \to y'$ in $\mathbf{D}$. Then $\pi_1 : \mathbf{P}_0 \to \mathbf{C}_0$ extends to a continuous functor $\mathbf{P} \to \mathbf{C}$, sending such an arrow β to the unique arrow $\delta : x \to x'$ for which

$\alpha' \circ \beta = \varphi(\delta) \circ \alpha$. This functor $\pi_1 : \mathbf{P} \to \mathbf{C}$ has a section $\sigma : \mathbf{C} \to \mathbf{P}$, sending an object $x \in \mathbf{C}_0$ to $(x, 1_{\varphi(x)} : \varphi(x) \tilde{\to} \varphi(x))$, and defined in the obvious way for arrows. Thus we have continuous functors

$$\mathbf{C} \underset{\sigma}{\overset{\pi_1}{\leftrightarrows}} \mathbf{P} \xrightarrow{s\pi_2} \mathbf{D}$$

where $s\pi_2\sigma = \varphi$ and $\pi_2\sigma = id_{\mathbf{C}}$. Furthermore, there is an evident continuous natural isomorphism $\theta : id_{\mathbf{P}} \to \sigma\pi_1$, with components

$$\theta_{x,\alpha} \; : \; y \tilde{\to} \varphi(x) \; = \alpha \; .$$

Thus π_1 is a categorical equivalence, and hence includes an equivalence of topoi $\mathcal{B}\mathbf{P} \cong \mathcal{B}\mathbf{C}$. It now suffices to prove that $s\pi_2 : \mathbf{P} \to \mathbf{D}$ induces an equivalence of topoi as well. In other words, since $s\pi_2$ is assumed to be proper or open, it remains to prove the proposition in the special case where $\varphi : \mathbf{C}_0 \to \mathbf{D}_0$ is itself a proper or open surjection, and we will return to this notation. To construct an inverse for the functor $\varphi^* : \mathcal{B}\mathbf{D} \to \mathcal{B}\mathbf{C}$, consider a $\mathbf{C}$-sheaf $S \to \mathbf{C}_0$. If x, x' are two points in $\mathbf{C}_0$ with $\varphi(x) = \varphi(x')$, there is a unique arrow in $\mathbf{C}$, denoted $\nabla_{x,x'} : x \to x'$, for which $\varphi(\nabla_{x,x'}) = 1_{\varphi}(x)$. This gives a map

$$\nabla \; : \; \mathbf{C}_0 \times_{\mathbf{D}_0} \mathbf{C}_0 \; \to \; \mathbf{C}_1 \; .$$

Pulling back the action by $\mathbf{C}$ on S along ∇ thus equips the sheaf $S \to \mathbf{C}_0$ with descent data (see Example 3.1(e)) for the map $\varphi : \mathbf{C}_0 \to \mathbf{D}_0$. Since this map φ is of effective descent, there is a sheaf T on $\mathbf{D}_0$, unique up to isomorphism, for which there is an isomorphism $u : \varphi^*(T) \cong S$ of sheaves on $\mathbf{C}_0$, compatible with descent data. It is now straightforward to descend the action by $\mathbf{C}$ on S to an action by $\mathbf{D}$ on T, in such a way that u is actually an isomorphism $\varphi^*(T) \cong S$ of $\mathbf{C}$-sheaves. This construction, of T out of S, provides an inverse (up to isomorphism) for the functor $\varphi^* : \mathcal{B}\mathbf{D} \to \mathcal{B}\mathbf{C}$.

§4 Diaconescu's theorem for s-étale categories

A topological category $\mathbf{C}$ is said to be *s-étale* if its source map $s : \mathbf{C}_1 \to \mathbf{C}_0$ is an étale map, i.e. a local homeomorphism. The modest purpose in this section is to extend the correspondence of Theorem 2.2 ("Diaconescu's theorem") to such s-étale topological categories. Note that it cannot possibly hold for the classifying topos of an arbitrary topological category, since such a topos may be degenerate (cf. Example 3.1(d)).

A $\mathbf{C}$-bundle on a space X is a sheaf $p : E \to X$, equipped with a continuous fiberwise left $\mathbf{C}$-action, given by maps

$$\pi : E \to \mathbf{C}_0 \quad , \quad a : \mathbf{C}_1 \times_{\mathbf{C}_0} E \to E \; .$$

The map a is defined for all pairs (g, e) where $g \in \mathbf{C}_1$, $e \in E$ and $s(g) = \pi(e)$, and is denoted $a(g, e) = g \cdot e$. That a is an action is expressed by the usual identities

$1_c \cdot e = e$ and $g \cdot (h \cdot e) = (g \circ h) \cdot e$; that it is fiberwise means that $p(g \cdot e) = p(e)$. Such a $\mathbb{C}$-bundle is said to be *principal* if the three conditions of Section 2 hold. We repeat them here for convenience: For any point $x \in X$,

(i) The stalk E_x is non-empty.

(ii) For any two points $y \in E_x$ and $z \in E_x$, there are a $w \in E_x$, and arrows $\alpha : \pi(w) \to \pi(y)$ and $\beta : \pi(w) \to \pi(z)$, such that $\alpha \cdot w = y$ and $\beta \cdot w = z$.

(iii) For any point $y \in E_x$, and any pair of arrows α, β in $\mathbb{C}$ with $s(\alpha) = \pi(y) = s(\beta)$ and $\alpha \cdot y = \beta \cdot y$, there exists a point $w \in E_x$ and an arrow $\gamma : \pi(w) \to \pi(y)$ in $\mathbb{C}$ such that $\gamma \cdot w = y$ in E_x and $\alpha\gamma = \beta\gamma$ in $\mathbb{C}$.

With the obvious notion of action preserving map, these principal $\mathbb{C}$-bundles over X form a category denoted

$$\mathrm{Prin}(X, \mathbb{C}) .$$

4.1. Theorem. *For any topological space X and any s-étale category $\mathbb{C}$, there is a natural equivalence*

$$\mathrm{Hom}(X, \mathcal{B}\mathbb{C}) \;\cong\; \mathrm{Prin}(X, \mathbb{C}) .$$

Proof. The proof is analogous to that of Theorem 2.2. For the construction of a topos morphism $f : X \to \mathcal{B}\mathbb{C}$ out of a principal bundle E, we again use the "tensor-product" construction. Thus, for a $\mathbb{C}$-sheaf S (an object of $\mathcal{B}\mathbb{C}$), we can construct a sheaf $S \otimes_{\mathbb{C}} E$ on X, by factoring out the fibered product space $S \times_{\mathbb{C}_0} E$ by the equivalence relation generated by the identifications

$$(s \cdot \alpha, e) \;\sim\; (s, \alpha \cdot e) .$$

The equivalence class of a pair (s, e) will again be denoted by $s \otimes e$. There is a projection $S \otimes_{\mathbb{C}} E \to X$, sending each equivalence class $s \otimes e$ to the point $p(e) \in X$; it is well-defined on equivalence classes since the action by $\mathbb{C}$ on E is fiberwise. To see that $S \otimes_{\mathbb{C}} E$ is a sheaf on X, we must show that this projection $S \otimes_{\mathbb{C}} E \to X$ is an étale map. To this end, construct $S \otimes_{\mathbb{C}} E$ as a coequalizer

$$S \times_{\mathbb{C}_0} \mathbb{C}_1 \times_{\mathbb{C}_0} E \rightrightarrows S \times_{\mathbb{C}_0} E \twoheadrightarrow S \otimes_{\mathbb{C}} E . \tag{1}$$

The two parallel maps here are given by the two actions, and send a triple (s, α, e) to $(s{\cdot}\alpha, e)$ and $(s, \alpha{\cdot}e)$, respectively. Now $S \times_{\mathbb{C}_0} \mathbb{C}_1 \times_{\mathbb{C}_0} E$ and $S \times_{\mathbb{C}_0} E$ are both sheaves on X, via the composite projections $S \times_{\mathbb{C}_0} \mathbb{C}_1 \times_{\mathbb{C}_0} E \to E \to X$ and $S \times_{\mathbb{C}_0} E \to E \to X$; indeed, these projections are both étale, since étale maps are stable under pullback and composition, while the maps $E \to X$, $S \to \mathbb{C}_0$ and $s : \mathbb{C}_1 \to \mathbb{C}_0$ are all assumed étale. The two parallel maps in (1) are maps of sheaves on X, so their coequalizer $S \otimes_{\mathbb{C}} E$ is again a sheaf on X.

Thus we have constructed a functor

$$- \otimes_{\mathbb{C}} E : \; \mathcal{B}\mathbb{C} \to (\text{sheaves on } X) .$$

This functor evidently commutes with colimits. To see that it commutes with finite limits, note that for a $\mathbb{C}$-sheaf S and a point $x \in X$, we have

$$(S \otimes_{\mathbb{C}} E)_x = S \otimes_{\mathbb{C}} (E_x) = \varinjlim_{I_x^{op}} \tilde{S}_x ,$$

where I_x is the category with E_x as a space of objects and $\mathbb{C}_1 \times_{\mathbb{C}_0} E_x$ (= pullback along $s : \mathbb{C}_1 \to \mathbb{C}_0$) as space of arrows. This category has a discrete topology, because $p : E \to X$ and $s : \mathbb{C}_1 \to \mathbb{C}_0$ are both étale. Thus, exactly as for Theorem 2.2, I_x^{op} is a filtering category when E is principal, so that $S \mapsto (S \otimes_{\mathbb{C}} E)_x$ commutes with finite limits. Since this holds for each point $x \in X$, the functor $S \mapsto S \otimes_{\mathbb{C}} E$ also commutes with finite limits, hence is the inverse image of a topos morphism f, uniquely determined up to isomorphism.

For the converse construction, of a principal bundle E from a morphism $f : X \to \mathcal{B}\mathbb{C}$, observe first that the étale map $s : \mathbb{C}_1 \to \mathbb{C}_0$ has the structure of a $\mathbb{C}$-sheaf, with (right) action given by composition. Thus $s : \mathbb{C}_1 \to \mathbb{C}_0$ underlies an object of $\mathcal{B}\mathbb{C}$, which will be denoted by $\tilde{\mathbb{C}}$. The inverse image functor f^* of any morphism $f : X \to \mathcal{B}\mathbb{C}$ thus gives an induced sheaf on X, defined as

$$E^f = f^*(\tilde{\mathbb{C}}) .$$

4.2. Lemma. *This sheaf E^f on X has the structure of a principal $\mathbb{C}$-bundle.*

Proof. In the proof, we shall explicitly use the assumption that $\mathbb{C}_0$ is sober. (Recall that all spaces are assumed sober.) We will also use that, since $s : \mathbb{C}_1 \to \mathbb{C}_0$ is étale, for each open $U \subseteq \mathbb{C}_0$ the space $t^{-1}(U)$ is a $\mathbb{C}$-sheaf, with sheaf projection $s : t^{-1}(U) \to \mathbb{C}_0$ and action given by composition. These sheaves $t^{-1}(U)$ generate the topos $\mathcal{B}\mathbb{C}$ (cf. the proof of Proposition 3.2). Note that $\tilde{\mathbb{C}}$ is the maximal generator. In particular, for each U we have

$$f^*(t^{-1}(U)) \subseteq f^*(\tilde{\mathbb{C}}) = E^f .$$

For the proof of the lemma, we first define a projection $\pi : E^f \to \mathbb{C}_0$ and an action $\mathbb{C}_1 \times_{\mathbb{C}_0} E^f \to E^f$. For the construction of π, let $x \in X$ and let $e \in E^f_x = f^*(\tilde{\mathbb{C}})_x$ be any point in the stalk over x. Consider the family

$$N_y = \{U \subseteq \mathbb{C}_0 \mid U \text{ open}, \ e \in f^*(t^{-1}U)_x\} .$$

As subobjects of $\tilde{\mathbb{C}}$ in $\mathcal{B}\mathbb{C}$, the objects $t^{-1}(U)$ satisfy the identities

$$t^{-1}(U) \cap t^{-1}(V) = t^{-1}(U \cap V) \qquad t^{-1}(\bigcup U_i) = \bigcup t^{-1}(U_i) ,$$

for any open sets U, V, U_i in $\mathbb{C}_0$. Since f^* preserves colimits and finite limits, it follows that the collection N_y of open sets is closed under intersection, and has the property that for any family of open subsets $\{U_i\}$ with $\bigcup U_i \in N_y$, some U_i must already belong to N_y. Thus the set $K = \mathbb{C}_0 - \bigcup\{U \subseteq \mathbb{C}_0 \mid U \text{ open}, \ U \notin N_y\}$ is an irreducible closed set. Since $\mathbf{C}_0$ is sober, K has a unique generic point, which we call $\pi(e)$. Thus $\pi(e)$ is the unique point in $\mathbf{C}_0$ with the property that for each open $U \subseteq \mathbf{C}_0$,

$$\pi(e) \in U \text{ iff } e \in f^*(t^{-1}(U))_x \ . \tag{2}$$

This construction, for each point $e \in E^f$, defines a map $\pi : E^f \to \mathbf{C}_0$. This map is evidently continuous, since $\pi^{-1}(U) = f^*(t^{-1}U) \subseteq E^f$ by (2).

To define the fiberwise left action by $\mathbf{C}$ on E^f, consider again a point $x \in X$, another one $e \in E^f_x$, and an arrow $\alpha : s(\alpha) \to t(\alpha)$ in $\mathbf{C}$ so that $\pi(e) = s(\alpha)$. Since the source map $s : \mathbf{C}_1 \to \mathbf{C}_0$ is étale, there exists an open neighbourhood U of $\pi(e)$ in $\mathbf{C}_0$ and a section $\varphi_\alpha : U \to \mathbf{C}_1$ of s so that $\varphi_\alpha(\pi e) = \alpha$. Composition with φ_α then defines a map of $\mathbf{C}$-sheaves (an arrow in $\mathcal{B}\mathbf{C}$)

$$\bar{\varphi}_\alpha : t^{-1}(U) \to \tilde{\mathbf{C}} \ , \qquad \bar{\varphi}_\alpha(\beta) = \varphi_\alpha(t(\beta)) \circ \beta \ .$$

Define the action of α on e by

$$\alpha \cdot e = f^*(\bar{\varphi}_\alpha)(e) \ \in \ E^f_x \ . \tag{3}$$

This action is readily seen to be continuous in α and e.

It remains to be shown that this action $\mathbf{C}_1 \times_{\mathbf{C}_0} E^f \to E^f$ satisfies the conditions (i) - (iii) for being principal. For the first condition, observe that for the terminal object 1 in $\mathcal{B}\mathbf{C}$, the unique map $\tilde{\mathbf{C}} \to 1$ is an epimorphism. Since $f^* : \mathcal{B}\mathbf{C} \to Sh(X)$ preserves epimorphisms as well as the terminal object, the map $E^f \to 1$ must be epi in $Sh(X)$; or in other words, each stalk E^f_x is non-empty. For condition (ii), consider for any open set $U \subseteq \mathbf{C}_0$, and any two sections $\varphi, \psi : U \to \mathbf{C}_1$ of the source map $s : \mathbf{C}_1 \to \mathbf{C}_0$, the induced arrow in the topos $\mathcal{B}\mathbf{C}$,

$$(\bar{\varphi}, \bar{\psi}) \ : \ t^{-1}(U) \to \tilde{\mathbf{C}} \times \tilde{\mathbf{C}}$$

(the product on the right is that of $\mathbf{C}$-sheaves). These maps, for all open $U \subseteq \mathbf{C}_0$ and all pairs of sections φ, ψ, together form a surjective map

$$\sum_{U,\varphi,\psi} t^{-1}(U) \twoheadrightarrow \tilde{\mathbf{C}} \times \tilde{\mathbf{C}}$$

in $\mathcal{B}\mathbf{C}$. Since f^* preserves products, sums and epis, the induced map of sheaves on X,

$$\sum_{U,\varphi,\psi} f^*(t^{-1}U) \to E^f \times_X E^f$$

is again surjective. Thus, if $x \in X$, and $y, z \in E^f_x$ are two points in the stalk over x, there are such U, φ, ψ and a point $w \in f^*(t^{-1}U)_x$ for which

$$f^*(\bar{\varphi})(w) = y \quad \text{and} \quad f^*(\bar{\psi})(w) = z \ .$$

Now write $\alpha = \varphi(\pi w)$ and $\beta = \psi(\pi w)$. Then, by definition (3) of the action, $\varphi = \varphi_\alpha$ and $\psi = \varphi_\beta$ on a possibly smaller neighbourhood $U' \subseteq U$ of $\pi(w)$, and $\alpha \cdot w = y$ while

$\beta \cdot w = z$. This shows that the action on E^f satisfies the second condition for being principal.

The verification of the third condition is similar; we omit the details.

The lemma now being verified, we complete the proof of the theorem. It remains to be shown that the two constructions, of a morphism $f : X \to \mathcal{B}\mathbf{C}$ out of a principal bundle E by $f^*(S) = S \otimes_{\mathbf{C}} E$, and of a principal bundle E^f out of a morphism f by $E^f = f^*(\tilde{\mathbf{C}})$, are mutually inverse, up to natural isomorphism. For this, observe first that for any (principal) $\mathbf{C}$-bundle E over X, with structure map $\pi : E \to \mathbf{C}_0$ as before, and for any open subset $U \subseteq \mathbf{C}_0$, there is a natural isomorphism

$$\mu_U : t^{-1}(U) \otimes_{\mathbf{C}} E \xrightarrow{\sim} \pi^{-1}(U) \subseteq E\,, \tag{4}$$

defined for any arrow α with $t(\alpha) \in U$ and any point $y \in E$ with $\pi(y) = s(\alpha)$, by $\mu_U(\alpha \otimes y) = \alpha \cdot y$. In particular, for $U = \mathbf{C}_0$, this gives an isomorphism $\tilde{\mathbf{C}} \otimes_{\mathbf{C}} E \cong E$. This shows that, starting with a principal bundle E, and constructing a map $f : X \to \mathcal{B}\mathbf{C}$, the bundle E^f associated to f is isomorphic to the bundle we started with:

$$E^f = f^*(\tilde{\mathbf{C}}) = \tilde{\mathbf{C}} \otimes_{\mathbf{C}} E \cong E\,.$$

The other way round, starting with a map $f : X \to \mathcal{B}\mathbf{C}$, we need to construct a natural isomorphism $\eta : - \otimes_{\mathbf{C}} E^f \to f^*$ between functors from $\mathcal{B}\mathbf{C}$ into $Sh(X)$. For each $\mathbf{C}$-sheaf S, define the component

$$\eta_S : S \otimes_{\mathbf{C}} E^f \to f^*(S)$$

as follows: for $s \in S$, and $y \in E^f$ over some point $\pi(y) \in \mathbf{C}_0$, choose first a section $\sigma : U \to S$ of the sheaf $S \to \mathbf{C}_0$ through s. This section gives a map $\bar{\sigma} : t^{-1}(U) \to S$ in $\mathcal{B}\mathbf{C}$, defined by $\bar{\sigma}(\alpha) = \sigma(t(\alpha)) \cdot \alpha$. Define $\eta_S(s \otimes y) = f^*(\bar{\sigma})(y)$. This gives a well-defined natural transformation $\eta : - \otimes_{\mathbf{C}} E \to f^*$. For a generator $S = t^{-1}(U)$ of $\mathcal{B}\mathbf{C}$, the component $\eta_S : t^{-1}(U) \otimes_{\mathbf{C}} E^f \to f^*(t^{-1}U)$ is precisely the isomorphism μ_U in (4) above. Thus μ is an isomorphism when restricted to generators. Since μ is natural in S, it follows that μ_S is an isomorphism for each $\mathbf{C}$-sheaf S. This completes the proof of Theorem 4.1. (We will be more explicit about naturality of the equivalence in Remark 4.5 below.)

Writing $k_{\mathbf{C}}(X)$ for the collection of *concordance* classes of principal $\mathbf{C}$-bundles over X (exactly as for discrete categories in Section 2), the theorem yields the following immediate corollary, by passing to homotopy classes of maps:

4.3. Corollary. *For any topological space X and any s-étale topological category $\mathbf{C}$, there is a natural bijective correspondence*

$$[X, \mathcal{B}\mathbf{C}] \cong k_{\mathbf{C}}(X)\,.$$

4.4. Topological groupoids. A topological groupoid is a topological category $\mathbf{C}$ equipped with an additional operation $i : \mathbf{C}_1 \to \mathbf{C}_1$ giving for each arrow $\alpha : x \to y$

in $\mathbb{C}$ a two-sided inverse $i(\alpha) = \alpha^{-1} : y \to x$. For example, every topological group is a topological groupoid, with the one-point space as space of objects. A topological groupoid is s-étale iff *all* its structure maps (source, target, composition, units and inverse) are local homeomorphisms. Topological groupoids which are s-étale play a central role in the theory of foliations (see e.g Haefliger(1958), (1984)) and typically arise from germs of homeomorphisms (or diffeomorphisms). For example, for any topological space M there is an s-étale topological groupoid $\Gamma(M)$ with M as space of objects, and with as space of arrows the space of all germs of (local) homeomorphisms, with the sheaf topology. For an s-étale topological groupoid $\mathbb{C}$ and a space X, a $\mathbb{C}$-bundle $E \to X$ over X is *principal* precisely when the map

$$\mathbb{C}_1 \times_{\mathbb{C}_0} E \to E \times_X E, \qquad (\alpha, e) \mapsto (\alpha \cdot e, e)$$

is an isomorphism. Thus principal $\mathbb{C}$-bundles are exactly the "$\mathbb{C}$-structures" considered in Haefliger(1958), and the collection of isomorphism classes of such is usually denoted $H^1(X, \mathbb{C})$. Thus Theorem 4.1 in this case provides a bijection

$$\pi_0 \mathrm{Hom}(X, \mathcal{B}\mathbb{C}) \;\cong\; H^1(X, \mathbb{C}) \,.$$

We conclude this section with some remarks on Theorem 4.1.

4.5. Remark. Just as for discrete categories, the naturality of the equivalence in Theorem 4.1 can be expressed by a commutative (up to isomorphism) diagram, for any map $f : Y \to X$ between spaces and any functor $\varphi : \mathbb{C} \to \mathbb{D}$ between s-étale topological categories:

$$\begin{array}{ccc}
\mathrm{Hom}(Y, \mathcal{B}\mathbb{C}) & \xrightarrow{\sim} & \mathrm{Prin}(Y, \mathbb{C}) \\
\uparrow & & \uparrow f^* \\
\mathrm{Hom}(X, \mathcal{B}\mathbb{C}) & \xrightarrow{\sim} & \mathrm{Prin}(X, \mathbb{C}) \\
\downarrow & & \downarrow \varphi_! \\
\mathrm{Hom}(X, \mathcal{B}\mathbb{D}) & \xrightarrow{\sim} & \mathrm{Prin}(X, \mathbb{D}).
\end{array}$$

The vertical maps on the left are again defined by composition, while f^* denotes the pullback of bundles. The covariant operation $\varphi_! : \mathrm{Prin}(X, \mathbb{C}) \to \mathrm{Prin}(X, \mathbb{D})$ on principal bundles is defined as follows. Denote by Φ the space $\mathbb{C}_0 \times_{\mathbb{D}_0} \mathbb{D}_1$ of pairs (x, β) where $x \in \mathbb{C}_0$ and $\beta \in \mathbb{D}_1$ is an arrow $\beta : \varphi(x) \to y$, with $s(\beta) = \varphi(x)$. Since $s : \mathbb{D}_1 \to \mathbb{D}_0$ is (assumed) étale, so is the projection $p_1 : \mathbb{C}_0 \times_{\mathbb{D}_0} \mathbb{D}_1 \to \mathbb{C}_0$. Furthermore, for an arrow $\alpha : x' \to x$ in $\mathbb{C}_1$, the action

$$(x, \beta) \cdot \alpha \;=\; (x', \beta \circ \varphi(\alpha))$$

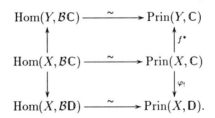

gives Φ the structure of a $\mathbb{C}$-sheaf, i.e. an object of $\mathcal{B}\mathbb{C}$. For the topos morphism $\varphi : \mathcal{B}\mathbb{C} \to \mathcal{B}\mathbb{D}$ induced by $\varphi : \mathbb{C} \to \mathbb{D}$, this $\mathbb{C}$-sheaf Φ is exactly the inverse image $\varphi^*(\tilde{\mathbb{D}})$ of the object $\tilde{\mathbb{D}}$ considered in the proof of Theorem 4.1. For the map $p_2 : \Phi \to \mathbb{D}_0$, sending a pair (x,β) to $y = t(\beta)$, the space Φ also carries a left action by the category $\mathbf{D}$, given by composition: for an arrow $\delta : y \to z$ in $\mathbf{D}$,

$$\delta \cdot (x,\beta) = (x, \delta \circ \beta) .$$

For a principal $\mathbf{C}$-bundle $E = (p : E \to X, \pi : E \to \mathbf{C}_0, \text{etc.})$ over X, define

$$\varphi_!(E) = \Phi \otimes_{\mathbf{C}} E .$$

Thus, a point of $\varphi_!(E)$ can be denoted $(x,\beta) \otimes e$ where $x \in \mathbf{C}_0, \beta : \varphi(x) \to y$ in $\mathbf{D}$ and $e \in E$ with $\pi(e) = x$. This space $\varphi_!(E)$ has a left $\mathbf{D}$-action, induced from that on Φ, and a natural projection $\varphi_!(E) \to X$, induced from $p : E \to X$. In this way, $\varphi_!(E)$ becomes a principal $\mathbf{D}$-bundle over X, and this is exactly the bundle corresponding to the composite map $X \to \mathcal{B}\mathbf{D}$. Indeed, write $h : X \to \mathcal{B}\mathbf{C}$ for the map corresponding to E under the equivalence of Theorem 4.1, so that $h^*(S) = S \otimes_{\mathbf{C}} E$ for any $\mathbf{C}$-sheaf S. The principal $\mathbf{D}$-bundle $E^{\varphi h}$, corresponding to the composite $\varphi \circ h : X \to \mathcal{B}\mathbf{D}$ under the equivalence, is constructed as

$$\begin{aligned} E^{\varphi h} &= (\varphi h)^*(\tilde{\mathbf{D}}) \\ &\cong h^* \, \varphi^*(\tilde{\mathbf{D}}) \\ &\cong h^*(\Phi) \\ &\cong \Phi \otimes_{\mathbf{C}} E \\ &= \varphi_!(E) . \end{aligned}$$

4.6 Remark. Let $\mathbf{C}$ be an s-étale topological category. It is a consequence of Theorem 4.1 that for any topological category $\mathbf{D}$ there is an equivalence of categories between topos morphisms $\mathcal{B}\mathbf{D} \to \mathcal{B}\mathbf{C}$ and principal $\mathbf{C}$-bundles over $\mathbf{D}$. These are principal $\mathbf{C}$-bundles over the space $\mathbf{D}_0$ of objects in the sense of Theorem 4.1, with an additional right action by $\mathbf{D}$, compatible with the principal left $\mathbf{C}$-action. (More generally, for any topos $\mathcal{E}$, there is an equivalence between maps $\mathcal{E} \to \mathcal{B}\mathbf{C}$ and principal $\mathbf{C}$-bundles over $\mathcal{E}$. In other words, X in Theorem 4.1 can be any topos. We will not use this more general result.)

§5 Sheaves on simplicial spaces

Recall that the simplicial model category Δ has as objects finite non-empty sets $[n] = \{0, \cdots, n\}$ (for $n \geq 0$), and as arrows $\alpha : [n] \to [m]$ monotone functions ($\alpha(i) \leq \alpha(j)$ whenever $i \leq j$). A simplicial space Y is a contravariant functor from Δ into spaces. Its value $Y([n])$ is denoted Y_n, and its action on an arrow α as above by $Y(\alpha) : Y_m \to Y_n$. In Deligne(1975), a sheaf S on Y is defined to be a system of sheaves S^n on Y_n (for $n \neq 0$), together with sheaf maps $S(\alpha) : Y(\alpha)^* S^n \to S^m$ for each $\alpha : [n] \to [m]$. These maps are required to satisfy the usual functoriality conditions: $S(\mathrm{id}) = \mathrm{id}$, and for $\alpha : [n] \to [m]$ and $\beta : [m] \to [k]$, the diagram

$$\begin{array}{ccc}
Y(\beta)^*\, Y(\alpha)^*\, (S^n) & \xrightarrow{Y(\beta)^*\, S(\alpha)} & Y(\beta)^*\, (S^m) \\
\Vert \wr & & \downarrow S(\beta) \\
Y(\beta\alpha)^*\, (S^n) & \xrightarrow{S(\beta\alpha)} & S^k
\end{array}$$

commutes. A morphism $f : S \to T$ between such sheaves consists of maps $f^n : S^n \to T^n$ of sheaves on Y_n, for each $n \geq 0$, which are compatible with the structure maps $S(\alpha)$ and $T(\alpha)$. This defines a category of sheaves on the simplicial space Y, which we denote by

$$Sh(Y)\,.$$

This category of sheaves is a topos; cf. Proposition 5.1 below.

Many important applications arise in the special case where Y is the nerve of a topological category $\mathbb{C}$. This is the simplicial space Nerve($\mathbb{C}$) with space of n-simplices Nerve$(\mathbb{C})_n$ the fibered product space $\mathbb{C}_1 \times_{\mathbb{C}_0} \times \cdots \times_{\mathbb{C}_0} \mathbb{C}_1$ of all composable strings of arrows $(x_0 \overset{\alpha_1}{\leftarrow} x_1 \leftarrow \cdots \overset{\alpha_n}{\leftarrow} x_n)$; for $n = 0$ this is just the space $\mathbb{C}_0$ of objects. For this particular case, the topos $Sh(\mathrm{Nerve}(\mathbb{C}))$ of sheaves will be denoted by $\mathcal{D}\mathbb{C}$, and referred to as the *Deligne classifying topos* of $\mathbb{C}$. Recall that for a general topological category $\mathbb{C}$, the more naive and much "smaller" classifying topos $\mathcal{B}\mathbb{C}$ described in Section 3 may contain no information about $\mathbb{C}$ (Example 3.1 (d)). It does, of course, when $\mathbb{C}$ is s-étale, as can be seen from the classification theorem 4.1. When $\mathbb{C}$ is not s-étale, $\mathcal{D}\mathbb{C}$ takes the role of $\mathcal{B}\mathbb{C}$. In Theorem 7.5 we will construct a weak homotopy equivalence $\mathcal{D}\mathbb{C} \to \mathcal{B}\mathbb{C}$ for any s-étale category $\mathbb{C}$.

The construction of the category $Sh(Y)$ is a special case of the more general context of a diagram of spaces Y indexed by some small category $\mathbf{K}$, i.e. a covariant functor Y from $\mathbf{K}$ into spaces. For an object $k \in \mathbf{K}$ we denote the value of Y at k by Y_k; the value of an arrow $\alpha : k \to \ell$ is denoted $Y(\alpha) : Y_k \to Y_\ell$. A sheaf on the diagram Y is defined to be a system of sheaves S^k on Y_k (for each object $k \in \mathbf{K}$), together with morphisms of sheaves $S(\alpha) : Y(\alpha)^*(S^\ell) \to S^k$ for each arrow $\alpha : k \to \ell$. With the evident morphisms, this defines a category $Sh(Y)$ of *sheaves on the diagram* Y.

From the category $\mathbf{K}$ and the diagram Y one can construct a topological category $Y_{\mathbf{K}}$: the objects of $Y_{\mathbf{K}}$ are the pairs (k, y) where $k \in \mathbf{K}$ and $y \in Y_k$; and an arrow $(k, y) \to (\ell, z)$ in $Y_{\mathbf{K}}$ is an arrow $\alpha : k \to \ell$ such that $Y(\alpha)(y) = z$. The topology on $Y_{\mathbf{K}}$ is that of the disjoint sum: its space of objects is $\Sigma_{k \in \mathbf{K}} Y_k$, and its space of arrows is $\Sigma_\alpha\, Y_{\mathrm{dom}(\alpha)}$ (where α ranges over all arrows in $\mathbf{K}$). The source map $s : \Sigma_\alpha\, Y_{\mathrm{dom}(\alpha)} \to \Sigma_k\, Y_k$ sends the summand $Y_{\mathrm{dom}(\alpha)}$ indexed by α to the summand Y_k, where $k = \mathrm{dom}(\alpha)$, via the identity map. The target map $t : \Sigma_\alpha\, Y_{\mathrm{dom}(\alpha)} \to \Sigma_k Y_k$ sends this summand $Y_{\mathrm{dom}(\alpha)}$ to Y_ℓ, where $\ell = \mathrm{cod}(\alpha)$, via the map $Y(\alpha)$. Notice that this topological category $Y_{\mathbf{K}}$ is evidently s-étale.

With this category $Y_{\mathbf{K}}$, the category $Sh(Y)$ of sheaves, as just defined, can be described as a classifying topos:

5.1. Proposition. *For any diagram Y of spaces, indexed by a small category K, there is a natural equivalence of topoi*

$$Sh(Y) \cong \mathcal{B}(Y_{\mathsf{K}}) .$$

Proof. This follows directly by a comparison of definitions.

Since Y_{K} is an s-étale category, we obtain:

5.2. Corollary. *For Y as in the previous proposition, and for any topological space X, there is a natural equivalence*

$$\mathrm{Hom}(X, Sh(Y)) \cong \mathrm{Prin}(X, Y_{\mathsf{K}}) ,$$

natural in X and Y.

The principal Y_{K}-bundles occurring in this corollary can be described in terms of principal K-bundles, in the following way. (We will continue to work here with a small (discrete) category K, but Proposition 5.3 below holds equally well for a topological category.) Recall that a principal K-bundle E over X consists of a system of sheaves E^k on X (one for each object $k \in \mathsf{K}$), and sheaf maps $E(\alpha) : E^k \to E^\ell$ for each $\alpha : k \to \ell$ in K, so that the principality conditions of Section 2 are satisfied. Call such a bundle E *augmented* (over Y) if E is equipped with a natural map aug: $E \to Y$ of diagrams of spaces. Thus aug is a system of maps $\mathrm{aug}^k : E^k \to Y_k$, so that for any arrow $\alpha : k \to \ell$ in K, the identity

$$Y(\alpha) \circ \mathrm{aug}^k = \mathrm{aug}^\ell \circ E(\alpha)$$

holds. With the obvious morphisms of principal bundles which respect the augmentation, one obtains a category

$$\mathrm{AugPrin}(X, \mathsf{K}, Y)$$

of principal K-bundles over X with an augmentation to Y.

5.3. Proposition. *For X and Y as above, there is a natural equivalence of categories*

$$\mathrm{Prin}(X, Y_{\mathsf{K}}) \cong \mathrm{AugPrin}(X, \mathsf{K}, Y)$$

Proof. Let E be a principal Y_{K}-bundle on X, with structure map

$$\pi : E \to (Y_{\mathsf{K}})_0 = \Sigma_k Y_k .$$

Then for each object $k \in \mathsf{K}$, the inverse image $E^k = \pi^{-1}(Y_k)$ is a sheaf on X equipped with a map $\pi^k : E^k \to Y_k$, defined as the restriction of π. Furthermore, if $\alpha : k \to \ell$ is an arrow in K, one can define a map $E(\alpha) : E^k \to E^\ell$ in terms of the given action by $Y_{\mathbf{K}}$ on E: for any point $e \in E^k$, there is an arrow $\tilde{\alpha} : (k, \pi(e)) \to (\ell, Y(\alpha)\pi(e))$ in $Y_{\mathbf{K}}$, and we define

$$E(\alpha)(e) = \tilde{\alpha} \cdot e .$$

It is elementary to verify that this augmented bundle E is again principal.

Conversely, from an augmented principal bundle F, one can define a $Y_{\mathbf{K}}$-bundle with underlying sheaf $E = \Sigma_{k \in \mathbf{K}} F^k$, structure maps $\pi : E \to Y = (Y_{\mathbf{K}})_0$ given by the augmentation $F^k \to Y_k (k \in \mathbf{K})$, and evident action by arrows in $Y_{\mathbf{K}}$. This action is again principal.

These two constructions provide the desired equivalence of categories.

We will examine this more closely in the special case of simplicial spaces. To this end, consider again the simplicial model category Δ, and its opposite Δ^{op}. It is well-known that the classifying topos $\mathcal{B}\Delta$ of simplicial sets "classifies" linear orders with end points (see Mac Lane-Moerdijk(1992), p. 463). A similar result holds for Δ^{op}, but without the endpoints. For a precise formulation (Proposition 5.4 below), define a *linear order* over a topological space X to be a sheaf L on X, together with a subsheaf $O \subseteq L \times_X L$, such that for each point $x \in X$ the stalk L_x is non-empty and linearly ordered by the relation

$$y \leq z \quad \text{iff} \quad (y, z) \in O_x \quad (\text{for } y, z \in L_x) .$$

A mapping $L \to L'$ between two such linear orders is a mapping of sheaves on X which for each point $x \in X$ restricts to an order-preserving map $L_x \to L'_x$ of stalks. This defines a category

$$Lin(X)$$

of linear orders over X. In the following proposition, using the notation of Section I.2, $\mathcal{B}(\Delta^{op})$ denotes the topos of presheaves on Δ^{op}, i.e. of *cosimplicial sets.*

5.4. Proposition. *For any topological space X, there is a natural equivalence of categories*

$$Hom(X, \mathcal{B}(\Delta^{op})) \cong Lin(X) .$$

Proof. By Theorem 2.2, there is a natural equivalence of categories, between topos maps $X \to \mathcal{B}(\Delta^{op})$ and principal Δ^{op}-bundles over X. Now a Δ^{op}-bundle over X is the same thing as a simplicial sheaf on X, and such a simplicial sheaf is principal whenever each stalk E_x is a principal simplicial set (i.e. a principal bundle over the one-point space). Following the definition given at the beginning of Section 2, a simplicial set S is principal iff it satisfies the following three conditions:

(i) S is non-empty;

(ii) given $y \in S_n$ and $z \in S_m$ there are arrows $\alpha : [n] \to [k]$ and $\beta : [m] \to [k]$ in Δ and a $w \in S_k$ so that $\alpha^*(w) = y$ and $\beta^*(w) = z$;

(iii) given $y \in S_n$ and $\alpha, \beta : [m] \rightrightarrows [n]$ in Δ with $\alpha^* y = \beta^* y$, there exists a $\gamma : [n] \to [k]$ in Δ and a $z \in S_k$ so that $\gamma^* z = y$ and $\gamma\alpha = \gamma\beta$.

Thus, the following lemma will complete the proof.

5.5. Lemma. *A simplicial set S is principal iff S is the nerve of a (uniquely determined) non-empty linear order.*

Proof. ($\Leftarrow$) For a linear order $(L, \leq)$, its nerve Nerve(L) is the simplicial set defined by

$$\mathrm{Nerve}_n(L) = \{(y_0, \cdots, y_n) \mid y_0 \leq \cdots \leq y_n \text{ in } L\} .$$

To show that this is a principal simplicial set, we verify the conditions (i)-(iii) just stated. Condition (i) holds since L is assumed non-empty. For condition (ii), suppose given sequences $y = (y_0, \cdots, y_n)$ and $z = (z_0, \cdots, z_m)$ in L. Let $k = n + m + 1$, and define $w = (w_0 \leq \cdots \leq w_k)$ to be the sequence made up from all the y_i and all the z_j, by putting them in the right order. Then there are strictly increasing functions $\alpha : [n] \to [k]$ and $\beta : [m] \to [k]$ so that $y_i = w_{\alpha(i)}$ and $z_j = w_{\beta(j)}$, for $i = 0, \cdots, n$ and $j = 0, \cdots, m$. Thus $y = \alpha^* w$ and $z = \beta^* w$, as required. For condition (iii), pick $y = (y_0, \cdots, y_n)$ and arrows $\alpha, \beta : [m] \rightrightarrows [n]$ in Δ with the property that $\alpha^* y = \beta^* y$. To find z and γ as in (iii) above, view y as a monotone map $y : [n] \to L$, and factor it as a surjection followed by an injection, say $\gamma : [n] \twoheadrightarrow [k]$ followed by $z : [k] \to L$. Thus $z \in \mathrm{Nerve}_k(L)$ and $y = \gamma^* z$. Furthermore, $\gamma\alpha = \gamma\beta$ since $\alpha^* y = \beta^* y$ while z is injective. This shows that Nerve(L) is a principal simplicial set.

($\Rightarrow$) Let S be a principal simplicial set. As pointed out in Example 2.1(d), S must preserve all finite limits which exist in Δ^{op}. Or in other words, S sends any finite colimit diagram in Δ to a finite limit diagram of sets. In particular, since any object $[n]$ in Δ can be constructed as a colimit by glueing copies of [1] together, viz. the colimit of

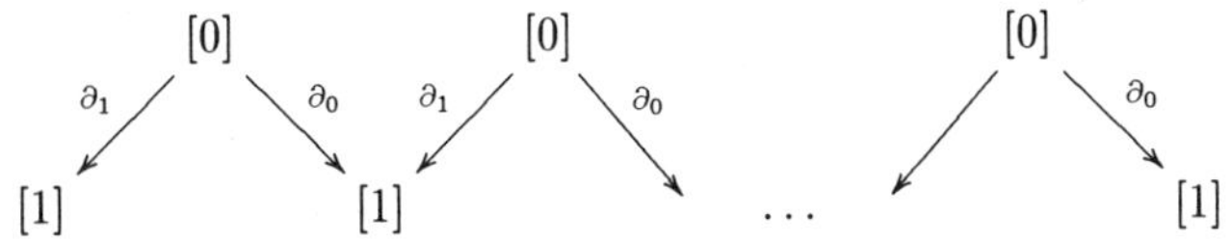

(n copies of [1]), it follows that there is a pullback

$$S_n = S_1 \times_{S_0} S_1 \times_{S_0} \cdots \times_{S_0} S_1 .$$

This means that S is the nerve of a category L.

Also, any principal functor preserves jointly monomorphic families. Or in other words, any surjective family

$$\{\alpha_i : [m_i] \to [n]\}_{i=1}^k$$

in Δ is sent to an injective function

$$(\alpha_1^*, \cdots, \alpha_k^*) : S_n \to S_{m_1} \times \cdots \times S_{m_k} .$$

In particular, when applied to the surjective family $\{\partial_0 : [0] \to [1], \partial_1 : [0] \to [1]\}$, this shows that $(d_0, d_1) : S_1 \to S_0 \times S_0$ is injective. Therefore, the category L, of which S

is the nerve, must be a preorder.

We now show that if S is principal then this preorder must be a non-empty linear order. First, L is non-empty since S is, by condition (i) for being principal. Next, to show that the order on L is total, pick $y, z \in L = S_0$. Since S is principal, there are arrows $\alpha : [0] \to [k]$ and $\beta : [0] \to [k]$ in Δ and a $w \in S_k$ so that $\alpha^* w = y$ and $\beta^* w = z$. Since S is the nerve of the preorder L, this means that $w = (w_o \leq \cdots \leq w_k)$ while $y = w_{\alpha(0)}$ and $z = w_{\beta(0)}$. Thus $y \leq z$ or $z \leq y$, according to whether $\alpha(0) \leq \beta(0)$ or $\beta(0) \leq \alpha(0)$. Finally, to show that the preorder is antisymmetric, suppose $y \leq z$ as well as $z \leq y$ in L. Thus $(y, z) \in S_1$ and $(z, y) \in S_1$, while $d_0(y, z) = d_1(z, y)$ and $d_1(y, z) = d_0(z, y)$. Since S is principal there must be a $k \geq 0$, and arrows $\alpha : [1] \to [k]$ and $\beta : [1] \to [k]$, and a $w \in S_k$, for which $(y, z) = \alpha^* w$ and $(z, y) = \beta^* w$ while $\partial_0 \alpha = \partial_1 \beta$ and $\partial_1 \alpha = \partial_0 \beta$. Thus $\alpha = \beta$, hence $y = z$.

This completes the proof of the lemma, and hence of Proposition 5.4.

Now let Y be a simplicial space, as in the beginning of this section. We will write

$$Lin(X, Y)$$

for the category of linear orders over X equipped with an augmentation into Y. Explicitly, if $L \to X$ is a linear order over X, then Nerve(L) is a simplicial sheaf on X, i.e. a simplicial space with étale maps into X. An augmentation of L into Y is a map of simplicial spaces aug : Nerve(L) $\to Y$. A morphism $(L, \text{aug}) \to (L', \text{aug}')$ in the category $Lin(X, Y)$ is a map $L \to L'$ of linear orders over X with the property that the induced map f : Nerve(L) $\to$ Nerve(L') of simplicial spaces respects the augmentations.

5.6. Corollary. *Let Y be a simplicial space. For any topological space X, there is a natural equivalence of categories*

$$\text{Hom}(X, Sh(Y)) \cong Lin(X, Y) .$$

Proof. This a special case of Proposition 5.3. Indeed, the simplicial space Y is a covariant functor on $\mathbf{K} = \Delta^{op}$, and $Sh(Y) = \mathcal{B}(Y_{\mathbf{K}})$ by Proposition 5.1. Thus, mappings $X \to Sh(Y)$ correspond by Corollary 5.2 and Proposition 5.3 to Y-augmented principal $\mathbf{K}$-bundles. By Proposition 5.4, these are precisely the Y-augmented linear orders over X.

A linear order can also be viewed as a topological category, with L as space of objects, and the order sheaf $O \subseteq L \times L$ as space of arrows. For any topological category $\mathbb{C}$, write $Lin(X, \mathbb{C})$ for the category of linear orders over X equipped with a continuous functor $L \to \mathbb{C}$. In terms of the notation $Lin(X, Y)$ of the previous corollary, this category $Lin(X, \mathbb{C})$ is just the same as $Lin(X, \text{Nerve}(\mathbb{C}))$. Thus, for the Deligne classifying topos $\mathcal{D}\mathbb{C}$, the previous corollary specializes to the following:

5.7. Corollary. *For any topological category* $\mathbb{C}$ *and any topological space* X, *there is a natural equivalence of categories*

$$\mathrm{Hom}(X, \mathcal{D}\mathbb{C}) \cong Lin(X, \mathbb{C}) .$$

For homotopy classes of maps, and the obvious notion of concordance, one obtains the following consequence, analogous to Corollary 4.3. Here $Lin_c(X, \mathbb{C})$ is the collection of concordance classes of objects from $Lin(X, \mathbb{C})$.

5.8. Corollary. *For* $\mathbb{C}$ *and* X *as in the previous corollary, there is a natural bijection*

$$[X, \mathcal{D}\mathbb{C}] \cong Lin_c(X, \mathbb{C}) .$$

§6 Cohomology of classifying topoi

This auxiliary section contains some remarks on the cohomology of the classifying topos $\mathcal{B}\mathbb{C}$ of a category $\mathbb{C}$, for later use in Chapter IV.

For a discrete category $\mathbb{C}$, it is well-known how to compute the cohomology of the topos $\mathcal{B}\mathbb{C}$ of presheaves on $\mathbb{C}$ (cf. Chapter I, Section 2). Let A be an abelian group in $\mathcal{B}\mathbb{C}$ (an object of $Ab(\mathcal{B}\mathbb{C})$, in the notation of Section I.4). Using the nerve of $\mathbb{C}$, one can define a cochain complex $C^{\cdot}(\mathbb{C}, A)$, with

$$C^n(\mathbb{C}, A) = \prod_{c_0 \leftarrow \ldots \leftarrow c_n} A(c_n);$$

the coboundary $d : C^{n-1}(\mathbb{C}, A) \to C^n(\mathbb{C}, A)$ is described as

$$(da)_{c_0 \xleftarrow{f_1} \ldots \xleftarrow{f_n} c_n} = \sum_{i=0}^{n-1} (-1)^i a_{d_i(c_0 \leftarrow \ldots \leftarrow c_n)} + (-1)^n A(f_n) a_{d_n(c_0 \leftarrow \ldots \leftarrow c_n)},$$

where $d_i(c_0 \leftarrow \ldots \leftarrow c_n)$ denotes the familiar simplicial boundary:

$$d_i(c_0 \xleftarrow{f_1} \ldots \xleftarrow{f_n} c_n) = \begin{cases} c_1 \leftarrow \ \ldots \ \leftarrow c_n & (i = 0) \\ c_0 \leftarrow \ \ldots \ c_{i-1} \xleftarrow{f_i \circ f_{i+1}} c_{i+1} \leftarrow \ldots \leftarrow c_n & (0 < i < n) \\ c_0 \leftarrow \ \ldots \ \leftarrow c_{n-1} & (i = n) . \end{cases}$$

The cohomology of this complex is (usually called) *the cohomology of the category* $\mathbb{C}$ with coefficients in A, and is denoted

$$H^{\cdot}(\mathbb{C}, A).$$

It is the same as the cohomology of the topos $\mathcal{B}\mathbb{C}$:

6.1. Proposition. *For any small category* $\mathbb{C}$, *and any abelian presheaf* A *on* $\mathbb{C}$, *there is a canonical isomorphism*

$$H^{\cdot}(\mathbb{C}, A) \cong H^{\cdot}(\mathcal{B}\mathbb{C}, A).$$

Proof. The proof is an immediate consequence of the existence of canonical projective resolutions in $\mathcal{B}\mathbb{C}$. We will use the following notation: for a presheaf of sets S (i.e. an object of $\mathcal{B}\mathbb{C}$), $\mathbb{Z}[S]$ denotes the free abelian presheaf on S; it can be constructed "pointwise", in the sense that $\mathbb{Z}[S](c)$ is the free abelian group on the set $S(c)$. For any abelian presheaf A, one has the usual adjunction formula

$$\mathrm{Hom}_{Ab(\mathcal{B}\mathbb{C})}(\mathbb{Z}[S], A) \cong \mathrm{Hom}_{\mathcal{B}\mathbb{C}}(S, A) \tag{1}$$

(where on the right, A is viewed as a presheaf of sets). In particular, for any object c in $\mathbb{C}$, (1) yields for the representable presheaf Yon(c) (cf. Section I.2) that

$$\mathrm{Hom}(\mathbb{Z}[\mathrm{Yon}(c)], A) \cong A(c). \tag{2}$$

In particular, $\mathbb{Z}[\mathrm{Yon}(c)]$ is projective. Now define a chain complex

$$\ldots. \to P_2 \to P_1 \to P_0 \to \mathbb{Z}$$

by

$$P_n = \sum_{c_0 \leftarrow \ldots \leftarrow c_n} \mathbb{Z}[\mathrm{Yon}(c_n)],$$

and with boundary maps $P_n \to P_{n-1}$ defined from the simplicial structure of the nerve of $\mathbb{C}$, in the usual way. For a fixed object c, the abelian group $P_n(c)$ is free on the set of all composable strings $c_0 \leftarrow \ldots \leftarrow c_n \leftarrow c$; so $P_\cdot(c)$ is the complex computing the simplicial homology (with integral coefficients) of the nerve of the category $c/\mathbb{C}$. Since $c/\mathbb{C}$ has an initial object, Nerve$(c/\mathbb{C})$ is contractible, hence $P_\cdot(c)$ is exact. This shows that $\cdots \to P_1 \to P_0 \to \mathbb{Z}$ is a projective resolution of $\mathbb{Z}$. Thus, for any abelian presheaf A on $\mathbb{C}$, the cohomology $H^\cdot(\mathcal{B}\mathbb{C}, A)$ can be computed using this projective resolution, as the cohomology of the complex $\mathrm{Hom}(P_\cdot, A)$. But, by (2) this is exactly the complex $C^\cdot(\mathbb{C}, A)$ described above.

6.2. Remark. In the special case for the category Δ^{op}, and the associated topos $\mathcal{B}(\Delta^{op})$ of cosimplicial sets, there is a much smaller projective resolution. For any $n \geq 0$, let P_n be free on the representable cosimplicial set $\mathrm{Yon}([n]) =\Delta([n], -)$. Define a boundary $\partial : P_n \to P_{n-1}$ as the alternating sum of the maps $\partial_i : P_n \to P_{n-1}$ induced by the maps "omit i": $[n-1] \hookrightarrow [n]$. For a fixed $k \geq 0$, the complex $P^k_\cdot = P_\cdot([k])$ computes the integral simplicial homology of the standard k-simplex, hence $P^k_\cdot$ is exact. Thus $P_\cdot$ is a projective resolution of $\mathbb{Z}$ in $\mathcal{B}(\Delta^{op})$. It follows that for any cosimplicial abelian group A, the groups $H^\cdot(\mathcal{B}\,\Delta^{op}, A)$ can be computed directly from the familiar complex $\cdots \to A^2 \to A^1 \to A^0$.

6.3. Remark. For a functor $\varphi : \mathbb{D} \to \mathbb{C}$ between categories, and the induced map $\varphi : \mathcal{B}\mathbb{D} \to \mathcal{B}\mathbb{C}$, the right derived functors $R^q\varphi_*$ can be described explicitly in the following well-known form. For an abelian presheaf A on $\mathbb{D}$ and any object $c \in \mathbb{C}$,

$$R^q\varphi_*(A)(c) = H^q(\varphi/c, A). \tag{3}$$

Here φ/c is the "comma category" with as objects the pairs $(d, u : \varphi d \to c)$ and as arrows $f : (d,u) \to (d',u')$ those $f : d \to d'$ in $\mathbb{D}$ for which $u' \circ \varphi(f) = u$. On the right of (3), A stands for the evident induced functor on φ/c, obtained from A by composition with the projection $\varphi/c \to \mathbb{D}$. Thus the Leray spectral sequence (Section I.4) for the map $\varphi : \mathcal{B}\mathbb{D} \to \mathcal{B}\mathbb{C}$ takes the form

$$E_2^{p,q} = H^p(\mathbb{C}, H^q(\varphi/-, A)) \Rightarrow H^{p+q}(\mathbb{D}, A).$$

After these introductory remarks, we will develop some analogues for topological categories. For an s-étale topological category $\mathbb{C}$, Proposition 6.1 takes the following form:

6.4. Proposition. *For any s-étale topological category* $\mathbb{C}$ *and for any abelian* $\mathbb{C}$*-sheaf* A*, there is a natural spectral sequence*

$$E_2^{p,q} = H^p H^q(\mathrm{Nerve}_\cdot(\mathbb{C}), \tilde{A}^\cdot) \Rightarrow H^{p+q}(\mathcal{B}\mathbb{C}, A).$$

Before embarking on the proof, we should explain the notation. Given an abelian group object A in $\mathcal{B}\mathbb{C}$, i.e. an abelian $\mathbb{C}$-sheaf, there is for each $n \geq 0$ a sheaf $\tilde{A}^n$ on $\mathrm{Nerve}_n(\mathbb{C})$, defined at the level of stalks by

$$(\tilde{A}^n)_\alpha = A_{x_n} \quad \text{for} \quad \alpha = \left(x_0 \overset{\alpha_1}{\leftarrow} x_1 \leftarrow \dots \overset{\alpha_n}{\leftarrow} x_n\right).$$

Furthermore, for a monotone map $\gamma : [n] \to [m]$ and the induced simplicial operator $\gamma^* : \mathrm{Nerve}_m(\mathbb{C}) \to \mathrm{Nerve}_n(\mathbb{C})$, the sheaf $\tilde{A}^n$ on $\mathrm{Nerve}_n(\mathbb{C})$ induces a sheaf $\gamma_!(\tilde{A}^n) := (\gamma^*)^*(\tilde{A}^n)$ on $\mathrm{Nerve}_m(\mathbb{C})$, related to $\tilde{A}^m$ via a homomorphism of abelian sheaves

$$\theta_\gamma : \gamma_!(\tilde{A}^n) \to \tilde{A}^m,$$

which is described at the level of stalks by using the action of $\mathbb{C}$ on A: for a point $\alpha = (x_0 \overset{\alpha_1}{\leftarrow} \dots \overset{x_m}{\leftarrow} x_m)$ in $\mathrm{Nerve}_m(\mathbb{C})$, the stalk of $\gamma_!(\tilde{A}^n)$ at α is $A_{x_{\gamma(n)}}$, and the action of the arrow $\alpha_{\gamma(n)+1} \circ \dots \circ \alpha_m$ on A gives a map $A_{x_{\gamma(n)}} \to A_{x_m}$, i.e. a map $\gamma_!(\tilde{A}^n)_\alpha \to (\tilde{A}^m)_\alpha$. This defines the stalk of θ_γ at the point $\alpha \in \mathrm{Nerve}_m(\mathbb{C})$. These maps θ_α make $n \mapsto H^q(\mathrm{Nerve}_n(\mathbb{C}), \tilde{A}^n)$ into a cosimplicial group, for each fixed $q \geq 0$; its cohomology makes up the E_2-term in 6.4.

Proof of 6.4. Define for each $n \geq 0$ a $\mathbb{C}$-sheaf D_n as follows: the total space of D_n is $\mathrm{Nerve}_{n+1}(\mathbb{C})$, the étale projection $p_n : D_n \to \mathbb{C}_0$ is defined by

$$p_n(x_0 \leftarrow \dots \leftarrow x_{n+1}) = x_{n+1},$$

and the action by $\mathbb{C}$ on D_n is defined by composition: for an arrow $\beta : y \to x_{n+1}$ in $\mathbb{C}$,

$$\left(x_0 \overset{\alpha_1}{\leftarrow} x_1 \leftarrow \dots \overset{\alpha_{n+1}}{\leftarrow} x_{n+1}\right) \cdot \beta = \left(x_0 \leftarrow \dots \leftarrow x_n \overset{\alpha_{n+1}\beta}{\leftarrow} y\right).$$

These sheaves D_n, for $n \geq 0$, together define a simplicial object $D.$ in the topos $\mathcal{B}\mathbb{C}$, with as stalk at a point $x \in \mathbb{C}_0$ the simplicial set $\mathrm{Nerve}(x/\mathbb{C})$. In particular, since the category $x/\mathbb{C}$ has an initial object, $D.$ is locally acyclic. Thus, by (4) of Section I.4 there is a standard spectral sequence

$$H^pH^q(\mathcal{B}\mathbb{C}/D_., \pi_.^*(A)) \Rightarrow H^{p+q}(\mathcal{B}\mathbb{C}, A), \tag{2}$$

where $\pi_n : \mathcal{B}(\mathbb{C})/D_n \to \mathcal{B}\mathbb{C}$ is the evident morphism of topoi (cf. the end of Section I.1). Next, there is a morphism of (simplicial) topoi

$$\lambda_n : \mathcal{B}\mathbb{C}/D_n \to Sh(\text{Nerve}_n(\mathbb{C})) \qquad \text{(all } n \geq 0),$$

with inverse image λ_n^* described as follows. Using the face map $d_{n+1} : D_n = \text{Nerve}_{n+1}(\mathbb{C}) \to \text{Nerve}_n(\mathbb{C})$, each object $F \in Sh(\text{Nerve}_n(\mathbb{C}))$, i.e. each étale map $p : F \to \text{Nerve}_n(\mathbb{C})$, gives an induced étale map $d_{n+1}^*(F) \to D_n$; when we equip $d_{n+1}^*(F)$ with the trivial $\mathbb{C}$-action, this étale map can be viewed as a map $d_{n+1}^*(F) \to D_n$ in $\mathcal{B}\mathbb{C}$. Define $\lambda_n^*(F)$ to be this last map, viewed as an object of $\mathcal{B}\mathbb{C}/D_n$. The direct image functor

$$\lambda_{n*} : \mathcal{B}\mathbb{C}/D_n \to Sh(\text{Nerve}_n(\mathbb{C}))$$

can be described explicitly at the level of stalks, for any object $w : E \to D_n$ of $\mathcal{B}\mathbb{C}/D_n$, by

$$\lambda_{n*}(w : E \to D_n)_{x_0 \leftarrow \ldots \leftarrow x_n} = w^{-1}(x_0 \leftarrow \ldots \leftarrow x_n \stackrel{id}{\leftarrow} x_n).$$

This functor λ_{n*} is evidently exact, hence induces an isomorphism in cohomology. Since $\lambda_{n*}\pi_n^*(A) = \tilde{A}^n$, this isomorphism takes the form

$$H^q(\mathcal{B}\mathbb{C}/D_n, \pi_n^*A) \cong H^q(\text{Nerve}_n(\mathbb{C}), \tilde{A}^n).$$

Using this isomorphism, the spectral sequence in (2) yields the one in the statement of the proposition.

Next, for a continuous functor $\varphi : \mathbb{D} \to \mathbb{C}$ into an s-étale category $\mathbb{C}$, we will write the Leray spectral sequence for the topos map $\mathcal{B}\mathbb{D} \to \mathcal{B}\mathbb{C}$ in a more explicit form, analogous to Remark 6.3. To this end, recall that any open subspace $U \subseteq \mathbb{C}_0$ gives rise to a $\mathbb{C}$-sheaf $s : t^{-1}(U) \to \mathbb{C}_0$, with action given by the composition operation of $\mathbb{C}$. For any other $\mathbb{C}$-sheaf S, there is a bijective correspondence between $\mathbb{C}$-equivariant maps $t^{-1}(U) \to S$ and sections of S over U:

$$\text{Hom}_{\mathcal{B}\mathbb{C}}(t^{-1}(U), S) \cong \Gamma(U, S).$$

For the topos map $\varphi : \mathcal{B}\mathbb{D} \to \mathcal{B}\mathbb{C}$, the pullback $\varphi^*(t^{-1}U)$ of this $\mathbb{C}$-sheaf will be denoted

$$U^{(\varphi)}.$$

So the points of $U^{(\varphi)}$ are pairs (α, x) where x is an object of $\mathbb{D}$ (a point in $\mathbb{D}_0$) and α is an arrow in $\mathbb{C}$ with $s\alpha = \varphi x$ and $t\alpha \in U$. This space $U^{(\varphi)}$ is made into a $\mathbb{D}$-sheaf via the étale projection $\pi : U^{(\varphi)} \to \mathbb{D}_0$ defined by $\pi(\alpha, x) = x$, and the action by $\mathbb{D}$ given by composition: $(\alpha, x) \cdot \beta = (\alpha \circ \varphi(\beta), s(\beta))$. It follows that the direct image functor $\varphi_* : \mathcal{B}\mathbf{D} \to \mathcal{B}\mathbf{C}$ can be described explicitly. For any $\mathbf{D}$-sheaf F, the sections of the sheaf $\varphi_*(F)$ over an open set $U \subseteq \mathbf{C}_0$ are given by

$$\Gamma(U, \varphi_* F) \cong \mathrm{Hom}_{\mathcal{B}\mathbf{D}}(U^{(\varphi)}, F). \tag{3}$$

An open set $U \subseteq \mathbf{C}_0$ not only gives rise to an object $U^{(\varphi)}$ of $\mathcal{B}\mathbf{D}$, but also to a topological category equipped with a functor into $\mathbf{D}$, together denoted

$$\pi_U : \varphi/U \to \mathbf{D}. \tag{4}$$

The objects of the category φ/U are the points of $U^{(\varphi)}$, i.e. the pairs (α, x) with $x \in \mathbf{D}_0$ and $\alpha : \varphi(x) \to y$ an arrow of $\mathbf{C}$ into some point $y \in U$. The arrows $\delta : (\alpha, x) \to (\alpha', x')$ in φ/U are arrows $\delta : x \to x'$ in $\mathbf{D}$ with the property that $\alpha' \circ \varphi(\delta) = \alpha$. The topology on φ/U is given by suitable fibered products: for the objects one has

$$(\varphi/U)_0 = U \times_{\mathbf{C}_0} \mathbf{C}_1 \times_{\mathbf{C}_0} \mathbf{D}_0,$$

while the space of arrows is topologized similarly as the pullback

$$(\varphi/U)_1 = U \times_{\mathbf{C}_0} \mathbf{C}_1 \times_{\mathbf{C}_0} \mathbf{D}_1.$$

Note that the projection functor π_U in (4), defined in the evident way, is étale (since the category $\mathbf{C}$ is assumed s-étale).

The object $U^{(\varphi)}$ of $\mathcal{B}\mathbf{D}$ and the category φ/U over $\mathbf{D}$ are related as follows:

6.5. Lemma. *There is a natural equivalence of topoi over $\mathcal{B}\mathbf{D}$:*

$$\mathcal{B}\mathbf{D}/U^{(\varphi)} \cong \mathcal{B}(\varphi/U).$$

Proof. Recall that the topos ("comma category") $\mathcal{B}\mathbf{D}/U^{(\varphi)}$ on the left hand side has as objects $\mathbf{D}$-sheaves F equipped with a $\mathbf{D}$-equivariant map $F \to U^{(\varphi)}$. Given an object E of $\mathcal{B}(\varphi/U)$, one obtains by composition $E \to (\varphi/U)_0 \xrightarrow{\pi_U} \mathbf{D}_0$ a sheaf on $\mathbf{D}_0$, with evident (right) action by $\mathbf{D}$, and $\mathbf{D}$-equivariant map $E \to U^{(\varphi)} = (\varphi/U)_0$. It is straightforward that this construction defines an equivalence of categories.

6.6. Proposition. *Let $\varphi : \mathbf{D} \to \mathbf{C}$ be a continuous functor into an s-étale category $\mathbf{C}$. Then for any abelian $\mathbf{D}$-sheaf A, the value $R^q\varphi_*(A)$ of the q-th right derived functor of $\varphi_* : \mathcal{B}\mathbf{D} \to \mathcal{B}\mathbf{C}$ is the sheaf associated to the presheaf on $\mathbf{C}_0$ defined for open sets $U \subseteq \mathbf{C}_0$ by*

$$U \mapsto H^q(\mathcal{B}(\varphi/U), \pi_U^*(A)).$$

Proof. Fix an integer $q \geq 0$ and an abelian $\mathbf{D}$-sheaf A, and define a presheaf $P^q(A)$ on the space $\mathbf{C}_0$ of objects of $\mathbf{C}$, as in the statement of the proposition:

$$P^q(A)(U) = H^q(\mathcal{B}(\varphi/U), \pi_U^* A),$$

for $U \subseteq \mathbf{C}_0$ open, where π_U is the projection functor $\varphi/U \to \mathbf{D}$ as in (4). An inclusion of open subsets $V \subseteq U$ induces a continuous functor $\varphi/V \to \varphi/U$, hence a morphism of topoi $\mathcal{B}(\varphi/V) \to \mathcal{B}(\varphi/U)$, hence a homomorphism of cohomology groups $H^q(\mathcal{B}(\varphi/U), \pi_U^* A) \to H^q(\mathcal{B}(\varphi/V), \pi_V^* A)$. This defines the restriction maps $P^q(A)(U) \to P^q(A)(V)$ for the presheaf $P^q(A)$. Denote the associated sheaf of this presheaf by $\tilde{P}^q(A)$. We claim that $\tilde{P}^q(A)$ carries a natural action by $\mathbf{C}$ from the right. Indeed, suppose $\alpha : x \to y$ is an arrow in $\mathbf{C}$, and V is any open neighbourhood of y. We can now use the assumption that $\mathbf{C}$ is s-étale, to find a neighbourhood $U \subseteq \mathbf{C}_0$ of x and a section $\sigma_\alpha : U \to \mathbf{C}$, of the source map, with $\sigma_\alpha(x) = \alpha$. If we choose U small enough, then $t \circ \sigma_\alpha : U \to \mathbf{C}_0$ will map U into the neighbourhood V of y. Then this section σ_α will induce a continuous functor $\mathrm{comp}(\sigma_\alpha) : \varphi/U \to \varphi/V$, simply by composition with the appropriate value of σ_α. This functor gives a morphism of topoi $\mathcal{B}(\varphi/U) \to \mathcal{B}(\varphi/V)$, and hence a homomorphism $H^q(\mathcal{B}(\varphi/V), \pi_V^* A) \to H^q(\mathcal{B}(\varphi/U), \pi_U^* A)$. This holds for all neighbourhoods V of y, naturally in V, so that one obtains a homomorphism

$$\alpha^* : \varinjlim_{y \in V} H^q(\mathcal{B}(\varphi/V), \pi_V^* A) \to \varinjlim_{x \in U} H^q(\mathcal{B}(\varphi/U), \pi_U^* A),$$

or in other words, a map of stalks

$$\alpha^* : \tilde{P}^q(A)_y \to \tilde{P}^q(A)_x.$$

This defines the action by $\mathbf{C}$ on $\tilde{P}^q(A)$, making $\tilde{P}^q(A)$ into a $\mathbf{C}$-sheaf. The construction is evidently functorial in A. The desired isomorphism

$$R^q \varphi_*(A) \cong \tilde{P}^q(A) \tag{5}$$

now follows by the uniqueness of right-derived functors: First, a short exact sequence $0 \to A \to B \to C \to 0$ induces for each open set $U \subseteq \mathbf{C}_0$ a long exact sequence

$$\ldots \to H^q(\mathcal{B}(\varphi/U), \pi_U^* A) \to H^q(\mathcal{B}(\varphi/U), \pi_U^* B) \to H^q(\mathcal{B}(\varphi/U), \pi_U^* C) \to \ldots,$$

hence a long exact sequence

$$\ldots \to \tilde{P}^q(A) \to \tilde{P}^q(B) \to \tilde{P}^q(C) \to \ldots.$$

Second, if A is an injective $\mathbf{D}$-sheaf, then for any object E of the topos $\mathcal{B}\mathbf{D}$, the cohomology groups $H^q(\mathcal{B}\mathbf{D}, E, A)$ (the values of the right-derived functors of $\mathrm{Hom}_{\mathcal{B}\mathbf{D}}(E, -)$) vanish for $q \geq 0$. But, by definition, these cohomology groups are those of the slice topos $\mathcal{B}\mathbf{D}/E$. Choosing $E = U^{(\varphi)}$ and using the equivalence of Lemma 6.5, we find that $H^q(\mathcal{B}(\varphi/U), \pi_U^* A) = 0$ whenever $q > 0$ and A is injective. It follows that $\tilde{P}^q(A) = 0$ for $q > 0$ and injective A. Finally it follows by (3) that $\tilde{P}^0$ is the functor φ_*. Thus, by uniqueness of derived functors, there is an isomorphism (5), natural in q and A. This proves the proposition.

Now let $\mathbf{K}$ be a discrete (small) category, and let Y be a $\mathbf{K}$-indexed diagram of spaces, as in Proposition 5.1. For the associated s-étale category $Y_{\mathbf{K}}$, there is an evident projection functor

$$\pi : Y_{\mathbf{K}} \to \mathbf{K},$$

and an associated topos map $\pi : \mathrm{Sh}(Y) = \mathcal{B}(Y_\mathbf{K}) \to \mathcal{B}\mathbf{K}$. For an abelian sheaf A on the diagram Y, the sheaf maps $A(\alpha) : Y(\alpha)^*(A^\ell) \to A^k$, for arrows $\alpha : k \to \ell$ in $\mathbf{K}$, induce homomorphisms of cohomology groups

$$H^*(Y_\ell, A^\ell) \to H^*(Y_k, A^k),$$

making $H^*(Y, A^\cdot)$ into a contravariant abelian group-valued functor on $\mathbf{K}$. As a special case of Proposition 6.6 we now obtain:

6.7. Corollary. *For any diagram Y of spaces on a small category $\mathbf{K}$, and for any abelian sheaf A on Y, there is a natural spectral sequence*

$$E_2^{p,q} = H^p(\mathbf{K}, H^q(Y, A^\cdot)) \Rightarrow H^{p+q}(\mathrm{Sh}(Y), A).$$

Proof. According to Proposition 6.6, it suffices to show for the projection $\pi :$ $Y_\mathbf{K} \to \mathbf{K}$ that

$$H^q(\mathcal{B}(\pi/k), \pi_k^*(A)) \cong H^q(Y_k, A^k). \tag{6}$$

Here π/k is the topological category with as objects pairs $(\alpha : \ell \to k, y)$ where α is an arrow in $\mathbf{K}$ and $y \in Y_\ell$. An arrow $(\alpha : \ell \to k, y) \to (\alpha' : \ell' \to k, y')$ in π/k is an arrow $\beta : (\ell, y) \to (\ell', y')$ in $Y_\mathbf{K}$, i.e. an arrow $\beta : \ell \to \ell'$ with $Y(\beta)(y) = y'$, such that $\alpha'\beta = \alpha$. The functor $\pi_k : \pi/k \to Y$ is the evident projection, sending $(\alpha : \ell \to k, y)$ to (ℓ, y). Consider the functors $i = i_k$ and $j = j_k$ in the diagram

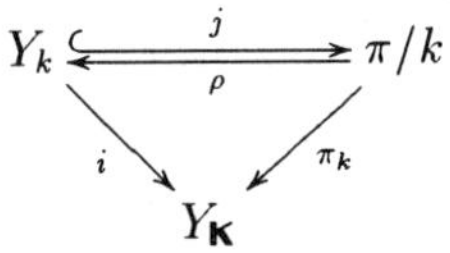

$$i(y) = (k, y)$$
$$j(y) = (\mathrm{id} : k \to k, y)$$

(where the space Y_k is viewed as a topological category with identity arrows only). For the associated topos morphisms

$$
\begin{array}{ccc}
\mathrm{Sh}(Y_k) & \xrightarrow{\quad j \quad} & \mathcal{B}(\pi/k) \\
& \searrow{\scriptstyle i} \quad \swarrow{\scriptstyle \pi_k} & \\
& \mathrm{Sh}(Y), &
\end{array}
$$

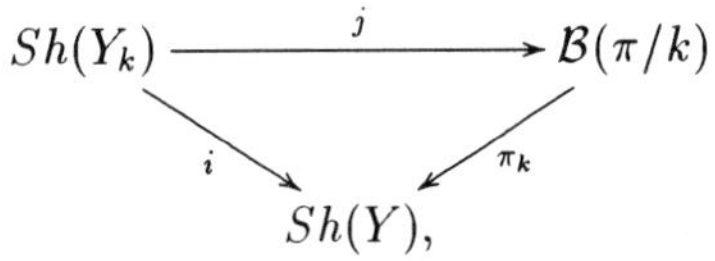

j induces an isomorphism

$$H^*(\mathcal{B}(\pi/k), B) \to H^*(Y_k, j^*B) \tag{7}$$

for any π/k-sheaf B. One way to see this uses the continuous functor $\rho : \pi/k \to Y_k$ defined on objects by $\rho(\alpha, y) = Y(\alpha)(y)$: for the induced topos map $\rho : \mathcal{B}(\pi/k) \to \mathrm{Sh}(Y_k)$, one has $j^* = \rho_*$. Thus j^* preserves injectives and commutes with the global sections functors. This gives the claimed isomorphism (7). For the special case where $B = \pi_k^*(A)$, it specializes to the desired isomorphism (6). This proves the corollary.

(The spectral sequence of Corollary 6.7 is analogous to the Bousfield-Kan spectral sequence for the diagram Y (Bousfield-Kan(1972)). In fact, for a locally constant sheaf A, the two correspond to each other via the isomorphism is cohomology between $Sh(Y)$ and the classifying space of $Y_{\mathbf{K}}$, provided by Theorem IV.2.1.)

For the special case where $\mathbf{K} = \Delta^{op}$, Remark 6.3 allows us to write the spectral sequence of 6.7 in a simpler form, involving for each $q \geq 0$ the cohomology of the simplicial group $H^q(Y_\cdot, A^\cdot)$:

6.8. Corollary. *For any simplicial space Y and any abelian sheaf A on Y, there is a natural spectral sequence*

$$E_2^{p,q} = H^p H^q(Y_\cdot, A^\cdot) \Rightarrow H^{p+q}(Sh(Y), A).$$

For later use, we also mention the following consequence of 6.7.

6.9. Corollary. *Let Y be a diagram of spaces on a small category $\mathbf{K}$, as before. If, for each $k \in \mathbf{K}$, the space Y_k is contractible, then $\pi : Sh(Y) \to \mathcal{B}\mathbf{K}$ is a weak equivalence of topoi.*

Proof. This follows from the toposophic Whitehead theorem (see Chapter I). Indeed, it is obvious that the functor π^* induces an equivalence between covering spaces of $\mathcal{B}\mathbf{K}$ and of $Sh(Y)$. Moreover, if A is any abelian group in $\mathcal{B}\mathbf{K}$, the spectral sequence of Corollary 6.7 collapses to an isomorphism $H^\cdot(\mathcal{B}\mathbf{K}, A) \tilde{\to} H^\cdot(Sh(Y), \pi^* A)$.

§7 Some homotopy equivalences between classifying topoi

In this section we will compare several topoi related to the classifying topos $\mathcal{B}\mathbb{C}$ of an s-étale category $\mathbb{C}$. First, for such a $\mathbb{C}$, we prove that there is a weak equivalence $\mathcal{D}\mathbb{C} \to \mathcal{B}\mathbb{C}$, comparing the Deligne classifying topos to $\mathcal{B}\mathbb{C}$ (Theorem 7.6 below). The other two comparisons (Propositions 7.7 and 7.8) are of a more technical nature, and will be used for our later comparison between classifying topoi and classifying spaces.

These homotopy equivalences involve a comparison of the fundamental groups, and we begin with a few remarks about this. Recall from Section 4 of Chapter I that the construction of the fundamental group and of the higher homotopy groups of a topos requires the topos to be locally connected. For a topos of the form $\mathcal{B}\mathbb{C}$, we have:

7.1. Lemma. *For any topological category $\mathbb{C}$ for which the spaces $\mathbb{C}_0$ and $\mathbb{C}_1$ of objects and arrows are both locally connected, the topos $\mathcal{B}\mathbb{C}$ is also locally connected.*

Proof. We need to show that under the assumptions of the lemma, the "constant functor" $\Delta : (sets) \to \mathcal{B}\mathbb{C}$, which to a set S associates the constant $\mathbb{C}$-sheaf $S \times \mathbb{C}_0 \to \mathbb{C}_0$ with the trivial action, has a left adjoint. This required left adjoint $\pi : \mathcal{B}\mathbb{C} \to (sets)$ is constructed as follows. For a $\mathbb{C}$-sheaf $E = (E, p : E \to \mathbb{C}_0,\ E \times_{\mathbb{C}_0} \mathbb{C}_1 \to E)$, both its structure map p and the pullback $E \times_{\mathbb{C}_0} \mathbb{C}_1 \to \mathbb{C}_1$ of it along $t : \mathbb{C}_1 \to \mathbb{C}_0$ are étale. So E and $E \times_{\mathbb{C}_0} \mathbb{C}_1$ are locally connected spaces since $\mathbb{C}_0$ and $\mathbb{C}_1$ are assumed to be. Let $\pi_0(E)$ and $\pi_0(E \times_{\mathbb{C}_0} \mathbb{C}_1)$ be their sets of connected components, and define $\pi(E)$ to be the coequalizer of the two maps $\pi_0(E \times_{\mathbb{C}_0} \mathbb{C}_1) \rightrightarrows \pi_0(E)$ induced by the projection and action maps $E \times_{\mathbb{C}_0} \mathbb{C}_1 \rightrightarrows E$. This construction, of the set $\pi(E)$ from the $\mathbb{C}$-sheaf E, defines the required left adjoint $\pi : \mathcal{B}\mathbb{C} \to (sets)$ for Δ.

We will call a topological category $\mathbb{C}$ *locally connected* if it satisfies the assumption in the statement of Lemma 7.1.

For any topological category $\mathbb{C}$ and its classifying topos $\mathcal{B}\mathbb{C}$, call a $\mathbb{C}$-sheaf E *invertible* if the action by arrows of $\mathbb{C}$ is invertible; this means that the map

$$E \times_{\mathbb{C}_0} \mathbb{C}_1 \to \mathbb{C}_1 \times_{\mathbb{C}_0} E \ , \quad (e, \alpha) \mapsto (\alpha, e \cdot \alpha)$$

is a homeomorphism. Let $\mathcal{I}\mathbb{C}$ be the full subcategory of $\mathcal{B}\mathbb{C}$ consisting of such invertible $\mathbb{C}$-sheaves. This category $\mathcal{I}\mathbb{C}$ is again a topos, and the inclusion $\mathcal{I}\mathbb{C} \hookrightarrow \mathcal{B}\mathbb{C}$ is the inverse image functor of a *connected* morphism of topoi

$$\theta : \ \mathcal{B}\mathbb{C} \to \mathcal{I}\mathbb{C}.$$

When $\mathbb{C}$ is a locally connected category, $\mathcal{I}\mathbb{C}$ is a locally connected topos: the proof of lemma 7.1 carries over verbatim to this case.

7.2. Lemma. *For any locally connected topological category $\mathbb{C}$, the morphism $\theta : \mathcal{B}\mathbb{C} \to \mathcal{I}\mathbb{C}$ induces an isomorphism of fundamental groups.*

Proof. It obviously suffices to prove that the inclusion functor $\theta^* : \mathcal{I}\mathbb{C} \to \mathcal{B}\mathbb{C}$ restricts to an equivalence of categories on locally constant objects (cf. Chapter I, Section 4). For this, we need to show that every locally constant object of $\mathcal{B}\mathbb{C}$ in fact belongs to the smaller category $\mathcal{I}\mathbb{C}$. Consider the Sierpinski-space Σ, with open point 1 and closed point 0. Let $Sh(\Sigma \times \mathbb{C}_1)$ be the topos of sheaves on the product space $\Sigma \times \mathbb{C}_1$. A sheaf F on $\Sigma \times \mathbb{C}_1$ is the same thing as a pair of sheaves F_0 and F_1 on $\mathbb{C}_1$, together with a map $u : F_0 \to F_1$. Thus there is a canonical morphism of topoi

$$\mu : Sh(\Sigma \times \mathbb{C}_1) \to \mathcal{B}\mathbb{C} \ ,$$

defined from the source and target maps $s, t : \mathbb{C}_1 \rightrightarrows \mathbb{C}_0$, by $\mu^*(E)_0 = t^*(E)$ and $\mu^*(E)_1 = s^*(E)$, and map $u : \mu^*(E)_0 \to \mu^*(E)_1$ given by the action of $\mathbb{C}$ on E. If E is locally constant, then so is the sheaf $\mu^*(E)$ on $\Sigma \times \mathbb{C}_1$, i.e. $\mu^*(E)$ is a covering space of $\Sigma \times \mathbb{C}_1$. By contractibility of Σ, it follows that $u : \mu^*(E)_0 \to \mu^*(E)_1$ must be an isomorphism. This says precisely that E belongs to $\mathcal{I}\mathbb{C}$. Thus every locally constant object of $\mathcal{B}\mathbb{C}$ belongs to $\mathcal{I}\mathbb{C}$, as required.

We now turn to the comparison between $\mathcal{D}\mathbb{C}$ and $\mathcal{B}\mathbb{C}$. Recall that the topos $\mathcal{D}\mathbb{C}$ of sheaves on the simplicial space Nerve($\mathbb{C}$) is the classifying topos of the s-étale category

$$\tilde{\mathbb{C}} = \text{Nerve}(\mathbb{C})_{\Delta^{op}}$$

(cf. Section 5). Explicitly, the objects of $\tilde{\mathbb{C}}$ are strings $\vec{\alpha} = (x_0 \overset{\alpha_1}{\leftarrow} x_1 \leftarrow \cdots \overset{\alpha_n}{\leftarrow} x_n)$ of composable arrows in $\tilde{\mathbb{C}}$; for two such strings $\vec{\alpha}$ and $\vec{\beta} = (y_0 \overset{\beta_1}{\leftarrow} y_1 \leftarrow \cdots \overset{\beta_m}{\leftarrow} y_m)$, an arrow $\vec{\alpha} \to \vec{\beta}$ in $\tilde{\mathbb{C}}$ is a morphism $\gamma : [m] \to [n]$ in the simplicial category Δ so that $\gamma^*(\vec{\alpha}) = \vec{\beta}$. There is an evident "last vertex" functor

$$\lambda : \tilde{\mathbb{C}} \to \mathbb{C}\,, \quad \lambda(x_0 \leftarrow \cdots \leftarrow x_n) = x_n.$$

This functor λ takes an arrow $\gamma : \vec{\alpha} \to \vec{\beta}$ as above to the composition $\alpha_{\gamma(m)+1} \circ \cdots \circ \alpha_m : x_n \to y_m = x_{\gamma(m)}$. The functor λ induces a morphism of topoi

$$\lambda : \mathcal{D}\mathbb{C} \to \mathcal{B}\mathbb{C}.$$

It is not difficult to see that its inverse image functor $\lambda^* : \mathcal{B}\mathbb{C} \to \mathcal{D}\mathbb{C}$ is full and faithful, so that (cf. Section I.4) $\lambda : \mathcal{D}\mathbb{C} \to \mathcal{B}\mathbb{C}$ is a *connected* morphism of topoi.

7.3. Lemma. *For any locally connected s-étale category $\mathbb{C}$, the map $\lambda : \mathcal{D}\mathbb{C} \to \mathcal{B}\mathbb{C}$ induces an isomorphism of fundamental groups.*

Proof. By Lemma 7.2, it clearly suffices to show that the restriction of λ^* : $\mathcal{B}\mathbb{C} \to \mathcal{B}\tilde{\mathbb{C}} = \mathcal{D}\mathbb{C}$ to invertible sheaves is an equivalence of categories

$$\lambda^* : \mathcal{I}\mathbb{C} \xrightarrow{\sim} \mathcal{I}\tilde{\mathbb{C}}.$$

To define an inverse for this functor λ^* on invertible sheaves, let T be a simplicial sheaf on Nerve($\mathbb{C}$), invertible when viewed as a $\tilde{\mathbb{C}}$-sheaf. Then its restriction T^0 to $\text{Nerve}_0(\mathbb{C}) = \mathbb{C}_0$ carries a natural action $T^0 \times_{\mathbb{C}_0} \mathbb{C}_1 \to T^0$ by $\mathbb{C}$, defined as the composite

$$T^0 \times_{\mathbb{C}_0} \mathbb{C}^1 = d_0^*(T^0) \xrightarrow{T(d_0)} T^1 \xrightarrow{T(d_1)^{-1}} d_1^*(T^0) \xrightarrow{p} T^0,$$

where p is the projection. This gives T^0 the structure of an invertible $\mathbb{C}$-sheaf, which we denote by $\lambda_!(T)$. It remains to observe that for any S in $\mathcal{I}\mathbb{C}$ and T in $\mathcal{I}\tilde{\mathbb{C}}$, there are natural isomorphisms $\lambda_!\lambda^*(S) \cong S$ and $\lambda^*\lambda_!(T) \cong T$. For example, for T, one has $\lambda^*\lambda_!(T)^n = (d_{n-1} \cdots d_0)^*(T^0)$, where $d_{n-1} \cdots d_0 : \text{Nerve}_n(\mathbb{C}) \to \text{Nerve}_0(\mathbb{C})$, and the isomorphism $\lambda^*\lambda_!(T)^n \to T^n$ is $T(d_{n-1} \cdots d_0) : (d_{n-1} \cdots d_0)^*(T^0) \to T^n$.

7.4. Lemma. *For any abelian $\mathbb{C}$-sheaf A, the map $\lambda : \mathcal{D}\mathbb{C} \to \mathcal{B}\mathbb{C}$ induces an isomorphism*

$$\lambda^* : H^n(\mathcal{B}\mathbb{C}, A) \xrightarrow{\sim} H^n(\mathcal{D}\mathbb{C}, \lambda^* A) \quad (n \geq 0).$$

Proof. This is immediate from the comparisons of the two spectral sequences in Proposition 6.4 and in Corollary 6.8, since the sheaf $\tilde{A}$ defined for Proposition 6.4 is precisely $\lambda^*(A)$.

From Lemmas 7.3 and 7.4 we deduce:

7.5. Theorem. *For any locally connected s-étale topological category* $\mathbb{C}$*, the map* $\lambda : \mathcal{D}\mathbb{C} \to \mathcal{B}\mathbb{C}$ *is a weak homotopy equivalence of topoi.*

Proof. This follows by the toposophic Whitehead theorem (Chapter I, Section 4), because λ induces an isomorphism in π_0 since λ is connected, in π_1 by Lemma 7.3 and in cohomology by Lemma 7.4.

The next two comparisons, in Propositions 7.6 and 7.7 below, are of an auxiliary nature. They will only be used in the proof of the comparison theorem 2.1 in Chapter IV.

To state the first, consider the category of simplices $\Delta(\mathbb{C})$ of a topological category $\mathbb{C}$. The objects of $\Delta(\mathbb{C})$ are strings $(x_0 \leftarrow \cdots \leftarrow x_n)$, and the arrows $(x_0 \leftarrow \cdots \leftarrow x_n) \to (y_0 \leftarrow \cdots \leftarrow y_m)$ are simplicial arrows $\gamma : [n] \to [m]$ so that $\gamma^*(y_0 \leftarrow \cdots \leftarrow y_m) = (x_0 \leftarrow \cdots \leftarrow x_n)$. Thus $\Delta(\mathbb{C})$ is the *dual* of the category $\tilde{\mathbb{C}}$ considered just above. For any topological category $\mathbb{C}$, the target map of the associated category $\Delta(\mathbb{C})$ is étale. Now let $\Delta_m(\mathbb{C}) \subseteq \Delta(\mathbb{C})$ be the topological category with the same space of objects as $\Delta(\mathbb{C})$, but with only those arrows given by *injective* $\gamma : [n] \to [m]$. This is of course again a t-étale category. There is a "first vertex" functor

$$\varphi : \Delta_m(\mathbb{C}) \to \mathbb{C}, \qquad \varphi(x_0 \leftarrow \cdots \leftarrow x_n) = x_0,$$

similar to the last vertex functor $\lambda : \tilde{\mathbb{C}} \to \mathbb{C}$. It induces a morphism of topoi $\varphi : \mathcal{B}(\Delta_m\mathbb{C}) \to \mathcal{B}\mathbb{C}$.

7.6. Proposition. *For any locally connected s-étale topological category* $\mathbb{C}$*, the map* $\varphi : \mathcal{B}(\Delta_m\mathbb{C}) \to \mathcal{B}\mathbb{C}$ *is a weak homotopy equivalence of topoi.*

Proof. Exactly as for the map $\lambda : \mathcal{D}\mathbb{C} \to \mathcal{B}\mathbb{C}$ in Lemma 7.3, it is easy to see that φ is a connected morphism, inducing an equivalence $\mathcal{I}(\Delta_m\mathbb{C}) \to \mathcal{I}\mathbb{C}$ of invertible sheaves and hence an isomorphism of fundamental groups. Thus, by the toposophic Whitehead theorem, it suffices to show that φ induces isomorphisms in cohomology with locally constant coefficients. To this end, consider the functor $\varphi : \Delta_m\mathbb{C} \to \mathbb{C}$ together with the identity functor $\iota : \mathbb{C} \to \mathbb{C}$, and for any open set $U \subseteq \mathbb{C}_0$ the associated categories φ/U and ι/U, as in (4) of Section 6. Also write U for the trivial topological category, with U as space of objects and with identity arrows only. There are continuous functors $t : \iota/U \rightleftarrows U : \eta$, where t is given by the target and $\eta(x)$ is the identity at x (in the category $\mathbb{C}$). Then $t \circ \eta$ is the identity functor on U, and ηt is related to the identity functor on ι/U via a continuous natural transformation. Thus, as explained in Section 3, the continuous functor $t : \iota/U \to U$ induces a weak homotopy equivalence of topoi $\mathcal{B}(\iota/U) \to \mathcal{B}U = Sh(U)$. (In fact $Sh(U)$ is a deformation retract of $\mathcal{B}(\iota/U)$.) Similarly, there are continuous functors

$$\overset{\tau}{\circlearrowright}\ \varphi/U \underset{\nu}{\overset{\varepsilon}{\rightleftarrows}} U$$

defined as follows: an object of φ/U is a pair $(\vec{\alpha}, \beta)$ where α is a string $\vec{\alpha} = (x_0 \xleftarrow{\alpha_1} x_1 \leftarrow \cdots \xleftarrow{\alpha_n} x_n)$ and β is an arrow $x_0 \to x$, into some point $x \in U$. The functor τ sends this pair to the augmented string $(x \xleftarrow{\beta} x_0 \xleftarrow{\alpha_1} \cdots \xleftarrow{\alpha_n} x_n)$ together with the identity arrow $x \to x$. The functor ε sends this pair $(\vec{\alpha}, \beta)$ simply to the point x. This definition of τ and ε on objects is extended in the evident way to arrows. Finally, the functor ν sends an object x to the pair $(x, u(x))$, consisting of the string x of length 0 and the identity arrow at x. Thus $\varepsilon \circ \nu$ is the identity functor, and there are natural transformations

$$id_{(\varphi/U)} \to \tau \leftarrow \nu \circ \varepsilon.$$

It follows that ε induces a (weak) homotopy equivalence $\mathcal{B}(\varphi/U) \to Sh(U)$ (cf. Section I.4). By combining these homotopy equivalences, we find that for a locally constant abelian group A in $\mathcal{B}\mathbf{C}$, Proposition 6.6 gives an isomorphism

$$R^q\varphi_*(\varphi^*A) \;\cong\; R^q\iota_*(\iota^*A) \quad (q \geq 0).$$

But ι is the identity functor, so $R^q\iota_* = 0$ for $q > 0$. Thus $\varphi : \mathcal{B}(\Delta_m\mathbf{C}) \to \mathcal{B}\mathbf{C}$ induces the required isomorphism in cohomology, and the proposition is proved.

Finally, for an s-étale category $\mathbf{C}$, we consider the following enlargement of the classifying topos $\mathcal{B}\mathbf{C}$, to be used for the comparison theorems in Chapter IV. Define a *quasi-*$\mathbf{C}$*-sheaf* S to be a $\mathbf{C}$-sheaf, except that the action $(s, \alpha) \mapsto s \cdot \alpha$ of $\mathbf{C}$ on S need *not* satisfy the identity law $s \cdot u(ps) = s$ (here $p : S \to \mathbf{C}_0$ is the structure map of S, and $u(ps)$ is the identity arrow at ps in $\mathbf{C}$). With the evident notion of action-preserving map between such quasi-$\mathbf{C}$-sheaves, these form a category denoted $\bar{\mathcal{B}}\mathbf{C}$. This category is a topos; in fact, $\bar{\mathcal{B}}\mathbf{C} = \mathcal{B}\mathbf{C}'$ where $\mathbf{C}'$ is obtained from $\mathbf{C}$ by adding "new" identity arrows: $\mathbf{C}'_1 = \mathbf{C}_1 + \mathbf{C}_0$. The evident full inclusion functor $\mathcal{B}\mathbf{C} \hookrightarrow \bar{\mathcal{B}}\mathbf{C}$ is the inverse image of a topos morphism

$$\psi : \bar{\mathcal{B}}\mathbf{C} \to \mathcal{B}\mathbf{C}.$$

For any quasi-$\mathbf{C}$-sheaf S, with sheaf projection $p : S \to \mathbf{C}_0$, one can construct a $\mathbf{C}$-sheaf $\tilde{S} \subseteq S$ by

$$\tilde{S} = \{s \in S \mid s \cdot u(ps) = s\}.$$

Since the "quasi-action" of $\mathbf{C}$ on S does satisfy the composition law ("$(s \cdot \alpha) \cdot \beta = s \cdot (\alpha\beta)$"), the identity

$$s \cdot u(ps) = (s \cdot u(ps)) \cdot u(ps)$$

holds for any point $s \in S$. Thus there is a retraction map

$$r = r_S : S \to \tilde{S} \ , \ r(s) = s \cdot u(ps).$$

The functor $\psi_* : \bar{\mathcal{B}}\mathbb{C} \to \mathcal{B}\mathbb{C}$ which sends S to $\tilde{S}$ is right adjoint to the inclusion functor $\psi^* : \mathcal{B}\mathbb{C} \to \bar{\mathcal{B}}\mathbb{C}$. (The unit of the adjunction is the identity map $T = \tilde{T}$ for any $\mathbb{C}$-sheaf T, and the counit is the inclusion $\tilde{S} \hookrightarrow S$ for any quasi-$\mathbb{C}$-sheaf S.) But, using the natural retraction r, it is clear that ψ_* is exact. In particular, ψ induces isomorphisms in cohomology $\psi^* : H^\cdot(\mathcal{B}\mathbb{C}, A) \xrightarrow{\sim} H^\cdot(\bar{\mathcal{B}}\mathbb{C}, \psi^* A)$ for any abelian $\mathbb{C}$-sheaf A. The morphism $\psi : \bar{\mathcal{B}}\mathbb{C} \to \mathcal{B}\mathbb{C}$ also induces isomorphisms in homotopy (when defined, i.e. when $\mathcal{B}\mathbb{C}$ and $\bar{\mathcal{B}}\mathbb{C}$ are locally connected).

In fact, the topos $\mathcal{B}\mathbb{C}$ is a natural deformation retract of $\bar{\mathcal{B}}\mathbb{C}$: the functor ψ_* is also left adjoint to the embedding $\mathcal{B}\mathbb{C} \hookrightarrow \bar{\mathcal{B}}\mathbb{C}$. (This time the counit $\psi_*\psi^*T \to T$ of the adjunction is the identity, while the unit $S \to \psi^*\psi_*S$ is the retraction map r.) Thus ψ_* is the inverse image of a morphism of topoi $\tau : \mathcal{B}\mathbb{C} \to \bar{\mathcal{B}}\mathbb{C}$, with $\psi\tau \cong id$ and $\tau\psi$ a natural retract of the identity functor. For later reference:

7.7. Proposition. *The classifying topos $\mathcal{B}\mathbb{C}$ of any topological category $\mathbb{C}$ is a natural deformation retract of the larger topos $\bar{\mathcal{B}}\mathbb{C}$ of quasi-$\mathbb{C}$-sheaves.*

Chapter III

Geometric Realization

§1 Geometric realization of simplicial spaces

The main purpose in this chapter will be to define a new geometric realization in the context of topoi, and show how the various classifying topoi, considered before, can be constructed in this way. As a motivation, and for later comparison, we begin by reviewing the well-known standard geometric realization for simplicial sets and for simplicial spaces.

Recall from Chapter II, Section 5, the simplicial model category $\mathbb{A}$ of finite ordered sets. A simplicial set is a presheaf on $\mathbb{A}$, i.e. a contravariant set-valued functor on $\mathbb{A}$. For such a simplicial set X, one writes X_n for $X([n])$ and $\alpha^* : X_n \to X_m$ for $X(\alpha)$, for any arrow $\alpha : [m] \to [n]$ in $\mathbb{A}$. As usual, d_i denotes the map $\partial_i^* : X_n \to X_{n-1}$, where $\partial_i : [n-1] \to [n]$ is the injective map which omits i from its range. Furthermore, Δ^n denotes the standard topological n-simplex (about which we shall have to be more explicit shortly). Each arrow $\alpha : [n] \to [m]$ gives an affine map $\Delta^n \to \Delta^m$, so as to make $\{\Delta^n : n \geq 0\}$ into a cosimplicial space (a functor from $\mathbb{A}$ into spaces). For a simplicial set X, its geometric realization $|X|$ is the topological space obtained from the disjoint sum $\sum_{n\geq 0} X_n \times \Delta^n$ (where X_n is given the discrete topology) by factoring out the equivalence relation generated by the identifications

$$(\alpha^*(x), t) \sim (x, \alpha(t)),$$

for any arrow $\alpha : [n] \to [m]$ in $\mathbb{A}$, any $x \in X_m$ and any $t \in \Delta^n$. In other words, $|X| = X_\cdot \otimes \Delta^\cdot$ is the tensor product of two functors from the category $\mathbb{A}$ into spaces, a covariant one $\Delta^\cdot$ and a contravariant one $X^\cdot$ (the latter taking values in discrete spaces). For basic properties of this geometric realization one may consult many standard sources, e.g. Milnor(1957), Gabriel-Zisman(1967), May(1967), or Fritsch-Piccinini(1990).

The same geometric realization can also be constructed by iterated adjunction spaces (pushouts): Let $X_k^{(nd)} \subseteq X_k$ be the set of non-degenerate k-simplices in X (those not in the image of α^* for a surjection $\alpha : [k] \to [m]$ where $m < k$). Then construct a sequence of spaces

$$|X|^{(0)} \subseteq |X|^{(1)} \subseteq |X|^{(2)} \subseteq \cdots$$

by induction, together with maps $\pi_k : X_k \times \Delta^k \to |X|^{(k)}$. By definition, $|X|^{(0)}$ is the set X_0 of vertices of X equipped with the discrete topology, and $\pi_0 : X_0 \times \Delta^0 \to |X|^{(0)}$ is the evident homeomorphism. Next, $|X|^{(k)}$ is constructed from $|X|^{(k-1)}$ and π_{k-1} as the pushout of topological spaces

$$\begin{array}{ccc} X_k^{(nd)} \times \partial\Delta^k & \hookrightarrow & X_k^{(nd)} \times \Delta^k \\ \downarrow & & \downarrow{\scriptstyle \pi_k} \\ |X|^{(k-1)} & \hookrightarrow & |X|^{(k)}. \end{array} \tag{1}$$

Here $\partial\Delta^k$ is the boundary of Δ^k, and X_k is given the discrete topology. From the map $\pi_{k-1} : X_{k-1} \times \Delta^{k-1} \to |X|^{(k-1)}$ one defines the map $X_k^{(nd)} \times \partial\Delta^k \to X^{(k-1)}$ on the left of (1): the restriction of the latter map to the i-th face $\Delta^{k-1} \hookrightarrow \partial\Delta^k$ is the composition $X_k \times \Delta^{k-1} \xrightarrow{d_i \times 1} X_{k-1} \times \Delta^{k-1} \xrightarrow{\pi_{k-1}} |X|^{(k-1)}$. The map $\pi_k : X_k \times \Delta^k \to |X|^{(k)}$ is defined by extending the map π_k on the right of (1) to degenerate simplices in the standard way. Now the geometric realization is defined as the union of the $|X|^{(k)}$, with the weak topology:

$$|X| = \bigcup_{k \geq 0} |X|^{(k)}.$$

We remark that, rather than restricting to the set $X_k^{(nd)}$ of non-degenerate k-simplices, one may also use the full set X_k in the construction of the iterated pushouts (1). The resulting bigger realization is well-known to be homotopy equivalent to the standard one $|X|$. For the realization of simplicial spaces below, we will only use this thicker realization.

For later purposes, it is necessary to be more explicit about a model for the standard n-simplex Δ^n. Let $I = [0,1] \subseteq \mathbf{R}$ be the unit interval, and define

$$\Delta^n = \{(x_1, \cdots, x_n) \mid x_i \in I,\ x_1 \leq \cdots \leq x_n\}.$$

The embedding of the i-th face $\partial^i : \Delta^{n-1} \to \Delta^n$, for $i = 0, \cdots, n$, is defined by

$$\partial^i(x_1, \cdots, x_{n-1}) = \begin{cases} (0, x_1, \cdots, x_{n-1}) & i = 0 \\ (x_1, \cdots, x_i, x_i, \cdots, x_{n-1}) & 0 < i < n \\ (x_1, \cdots, x_{n-1}, 1) & i = n. \end{cases}$$

Then the boundary of Δ^n is defined as

$$\partial\Delta^n = \bigcup_{i=0}^{n} \partial^i(\Delta^{n-1}) = \{(x_1, \cdots, x_n) \in \Delta^n | x_1 = 0 \text{ or } x_n = 1 \text{ or } \exists i(x_i = x_{i+1})\}.$$

1.1. Remark. It is important to note that we have only used the order relation $\leq$ on I and its endpoints $0, 1$. Thus any topological space J with an order $\leq$ and endpoints $0, 1$ (an "interval") gives rise to a cosimplicial space $\Delta^{\cdot}_{(J)}$ (a functor $k \mapsto \Delta^k_{(J)}$ from $\mathbb{\Delta}$ into spaces), and hence a realization for any simplicial set X:

$$|X|_{(J)} = X_{\cdot} \otimes \Delta^{\cdot}_{(J)}.$$

This tensor product can again be constructed by iterated pushouts, now of the form

$$\begin{array}{ccc} X_k^{(nd)} \times \partial\Delta^k_{(J)} & \hookrightarrow & X_k^{(nd)} \times \Delta^k_{(J)} \\ \downarrow & & \downarrow \\ |X|^{(k-1)}_{(J)} & \hookrightarrow & |X|^{(k)}_{(J)} , \end{array} \tag{2}$$

so that

$$|X|_{(J)} = \bigcup_{k \geq 0} |X|^{(k)}_{(J)} .$$

Note that this construction is functorial: If $\varphi : J \to J'$ is a continuous map of "intervals" (φ preserves the order-relation as well as the endpoints) then φ induces a natural continuous map

$$\varphi_X : |X|_{(J)} \to |X|_{(J')},$$

for any simplicial set X.

1.2. Example. Let $\Sigma = \{0, 1\}$ be the Sierpinski space, with a closed point 0 and an open point 1, and order $0 \leq 1$. Then

$$\begin{array}{rcl} \Delta^0_{(\Sigma)} & = & pt, \\ \Delta^1_{(\Sigma)} & = & \Sigma, \\ \partial\Delta^1_{(\Sigma)} & = & \{0,1\} \text{ with discrete topology}, \\ \partial\Delta^2_{(\Sigma)} & = & \Delta^2_{(\Sigma)} = \{(0,0),(0,1),(1,1)\}, \end{array}$$

and so on: $\partial\Delta^n_{(\Sigma)} = \Delta^n_{(\Sigma)}$ for $n \geq 2$. Thus, when one constructs the geometric realization of a simplicial set X with respect to the interval Σ, the series of pushouts (2) (for $J = \Sigma$) stops after one step:

$$X_0 = |X|^{(0)}_{(\Sigma)} \subseteq |X|^{(1)}_{(\Sigma)} = |X|_{(\Sigma)}.$$

In other words, the realization $|X|_{(\Sigma)}$ is the space $X_1 \times \Sigma$ factored out by the equivalence relation which identifies $(x, 0)$ and $(x, 1)$ if x is degenerate or if $d_0 x = d_1 x$. Surely this is not a very interesting realization.

Next, we should make some remarks concerning the geometric realization of simplicial spaces. For a simplicial space X, the notation $|X|$ will always denote the "thickened" geometric realization, as described in Segal(1974). Thus, $|X|$ is obtained from $\Sigma_{n \geq 0} X_n \times \Delta^n$ by making the identifications

$$(\alpha^*(x), t) \sim (x, \alpha(t)) \tag{3}$$

as for a simplicial set, but now only all *injective* order-preserving functions $\alpha : [n] \to [m]$, with associated maps $\alpha^* : X_m \to X_n$ and $\alpha : \Delta^n \to \Delta^m$, and for all $x \in X_m$ and $t \in \Delta^n$. This geometric realization has various well-known basic properties, described in the Appendix of Segal(1974). In particular, we mention the property that for a map $f : X \to Y$ between simplicial spaces, its realization $|f| : |X| \to |Y|$ is a weak homotopy equivalence whenever $f_n : X_n \to Y_n$ is (for each $n \geq 0$). Recall also that if all the degeneracies $s_i : X_{n-1} \to X_n$ of a simplicial space X are closed cofibrations, then the thickened realization $|X|$ is homotopy equivalent to the realization which is defined by the identifications (3) for all α, not just injective ones.

As before, one can build up this thickened realization by iterated pushouts, defining

$$|X|^{(0)} \subseteq |X|^{(1)} \subseteq |X|^{(2)} \subseteq \cdots$$

inductively, by $|X|^{(0)} = X_0$, and by the pushout diagram

$$\begin{array}{ccc} X_n \times \partial\Delta^n & \hookrightarrow & X_n \times \Delta^n \\ {\scriptstyle v_n}\downarrow & & \downarrow{\scriptstyle u_n} \\ |X|^{(n-1)} & \hookrightarrow & |X|^{(n)}, \end{array} \tag{4}$$

where the left-hand vertical map v_n is defined just as in the discrete case: on the i-th face $X_n \times \Delta^{n-1} \hookrightarrow X_n \times \partial\Delta^{n-1}$ it is the composite of $d_i \times 1 : X_n \times \Delta^{n-1} \to X_{n-1} \times \Delta^{n-1}$ and the map $u_{n-1} : X_{n-1} \times \Delta^{n-1} \to |X|^{(n-1)}$. (The "thickening" is reflected by the fact that, in (4), one does not just consider non-degenerate simplices in X_n, as in (1) above.) The thickened realization is now constructed as

$$|X| = \bigcup_{n \geq 0} |X|^{(n)}, \tag{5}$$

again with the weak topology.

§2 Classifying spaces

This section contains some remarks on the sheaf cohomology of classifying spaces. In accordance with the remarks in the preface, we will treat the case of discrete (small) categories first.

For a small category $\mathbb{C}$, its classifying space $B\mathbb{C}$ is the (ordinary) geometric realization of the simplicial set Nerve($\mathbb{C}$). We will give a "cellular" description of the cohomology groups $H^\cdot(B\mathbb{C}, A)$ of this classifying space $B\mathbb{C}$ with coefficients in any abelian sheaf A on $B\mathbb{C}$. To this end, first recall from Chapter II, Section 7 the category of simplices $\Delta\mathbb{C}$ associated to an arbitrary small category $\mathbb{C}$: The objects of $\Delta\mathbb{C}$ are pairs (n, α) where $n \geq 0$ and $\alpha = (x_0 \xleftarrow{\alpha_1} x_1 \leftarrow \cdots \xleftarrow{\alpha_n} x_n)$ is an n-simplex in the nerve of $\mathbb{C}$. An arrow $u : (n, \alpha) \to (m, \beta)$ in $\Delta\mathbb{C}$ is by definition an arrow $u : [n] \to [m]$ in $\mathbb{\Delta}$ with the property $\alpha = u^*(\beta)$. The "first vertex" functor

$$\varphi : \Delta\mathbb{C} \to \mathbb{C}$$

sends an object (n, α) to x_0, and an arrow $u : (n, \alpha) \to (m, \beta)$ to $\beta_1 \circ \cdots \circ \beta_{u(0)} : x_0 = y_{u(0)} \to y_0$ (where $\beta = (y_0 \leftarrow \cdots \leftarrow y_m)$). This functor $\varphi : \Delta\mathbb{C} \to \mathbb{C}$ is well-known to induce a homotopy equivalence of classifying spaces

$$B\varphi : B\Delta\mathbb{C} \to B\mathbb{C}.$$

Indeed, this follows immediately from Quillen's "Theorem A", since for any fixed object $x \in \mathbb{C}$ the category φ/x is contractible: in the notation of Quillen (1973), the natural transformations

$$(x_0 \leftarrow \cdots \leftarrow x_n, x_0 \to x) \xrightarrow{\partial_0} (x \leftarrow x_0 \leftarrow \cdots \leftarrow x_n, x \xrightarrow{id} x) \xrightarrow{\partial_1^{n+1}} (x, x \xrightarrow{id} x) \tag{1}$$

connect the identity functor to the constant functor on φ/x with value $(x, x \xrightarrow{id} x)$ (or more explicitly, with value $((0,x), x \xrightarrow{id} x)$).

We now associate to each abelian sheaf A on the classifying space $B\mathbb{C}$ a contravariant functor $\gamma(A)$ from $\Delta\mathbb{C}$ into abelian groups:

$$\gamma(A) : (\Delta\mathbb{C})^{op} \to \underline{Ab} \ .$$

For each object (n, α) of $\Delta\mathbb{C}$ as above, there is an associated map from a copy Δ^n_α of the standard n-simplex Δ^n into the classifying space, denoted

$$\pi_\alpha : \Delta^n_\alpha \to B\mathbb{C} \ .$$

Furthermore, if $u : (n, \alpha) \to (m, \beta)$ is an arrow in $\Delta\mathbb{C}$ then $u : [n] \to [m]$ induces an affine map (again denoted) $u : \Delta^n_\alpha \to \Delta^m_\beta$ between standard simplices, for which

$$\pi_\beta \circ u = \pi_\alpha \ .$$

Now define the functor $\gamma(A)$ on objects by

$$\gamma(A)(n, \alpha) = \Gamma(\Delta^n_\alpha, \ \pi^*_\alpha(A)) \ .$$

An arrow $u : (n, \alpha) \to (m, \beta)$ induces a homomorphism

$$\Gamma(\Delta^m_\beta, \ \pi^*_\beta(A)) \to \Gamma(\Delta^n_\alpha, \ u^*\pi^*_\beta A) \ \cong \ \Gamma(\Delta^n_\alpha, \ \pi^*_\alpha A) \ ,$$

and this defines $\gamma(A)$ on arrows.

This construction is of course functorial in A. In other words, denoting (as in Chapter I, Section 4) the category of abelian sheaves on $B\mathbb{C}$ by $Ab(Sh(B\mathbb{C}))$, and the category of abelian presheaves on $\Delta\mathbb{C}$ by $Ab(\mathcal{B}(\Delta\mathbb{C}))$, we have a functor

$$\gamma : Ab(Sh(B\mathbb{C})) \to Ab(\mathcal{B}(\Delta\mathbb{C})) \ . \tag{2}$$

This functor is evidently left-exact. Its right-derived functors $R^q\gamma$ are described by the following lemma.

2.1. Lemma. *For any $q \geq 0$ and any abelian sheaf A on $B\mathbb{C}$ there is a canonical isomorphism*

$$R^q\gamma(A)(n,\alpha) \cong H^q(\Delta^n_\alpha, \pi^*_\alpha(A)) .$$

Proof. Define for each abelian sheaf A on $B\mathbb{C}$ a functor $\mathcal{H}^q(A) : (\Delta\mathbb{C})^{op} \to \underline{Ab}$, by $\mathcal{H}^q(A)(n,\alpha) = H^q(\Delta^n_\alpha, \pi^*_\alpha(A))$. Then clearly a short exact sequence $0 \to A' \to A \to A'' \to 0$ of abelian sheaves induces a long exact sequence $\cdots \to \mathcal{H}^n(A') \to \mathcal{H}^n(A) \to \mathcal{H}^n(A'') \to \mathcal{H}^{n+1}(A') \to \cdots$. Furthermore, if A is an injective sheaf on $B\mathbb{C}$, its restriction to the closed subset $\pi_\alpha(\Delta^n_\alpha) \subseteq B\mathbb{C}$ is soft, so $H^q(\pi_\alpha(\Delta^n_\alpha), A) = 0$ for $q > 0$. Now $\pi_\alpha(\Delta^n_\alpha)$ is a possibly degenerate n-simplex, and $\pi_\alpha : \Delta^n_\alpha \to \pi_\alpha(\Delta^n_\alpha)$ is a proper map with contractible fibers, so (by "proper base-change", Godement (1958), p. 202) $H^q(\Delta^n_\alpha, \pi^*_\alpha(A)) \cong H^q(\pi_\alpha(\Delta^n_\alpha), A) \cong 0$. This shows that $\mathcal{H}^q(A) = 0$ whenever A is injective and $q > 0$. By uniqueness of derived functors, there is an isomorphism $\mathcal{H}^q(A) \cong R^q\gamma(A)$, natural in A. This proves the lemma.

Thus, for suitable sheaves A, the cohomology groups $H^q(B\mathbb{C}, A)$ of the classifying space $B\mathbb{C}$ can be computed in terms of the cohomology of the category $\Delta\mathbb{C}$ (cf. Chapter II, Section 6):

2.2. Corollary. *If A is a sheaf on the classifying space $B\mathbb{C}$ with the property that $H^q(\Delta^n_\alpha, \pi^*_\alpha(A)) = 0$ for each n-simplex $\Delta^n_\alpha \to B\mathbb{C}$, then there is a natural isomorphism*

$$H^q(B\mathbb{C}, A) \cong H^q(\Delta\mathbb{C}, \gamma A).$$

The corollary applies in particular when the sheaf A on $B\mathbb{C}$ is locally constant. It also applies to a slightly larger class of "pseudo-constant" sheaves, defined as follows. Let Δ^n be a standard n-simplex, and let $\partial_n : \Delta^{n-1} \hookrightarrow \Delta^n$ be the inclusion of the last face. By recursion on n, we first define a sheaf A on Δ^n to be pseudo-constant if A is constant on $\Delta^n - \partial_n(\Delta^{n-1})$, and if the restriction of A to this last face $\Delta^{n-1} \subseteq \Delta^n$ is a pseudo-constant sheaf on Δ^{n-1}. Then a sheaf A on $B\mathbb{C}$ is defined to be pseudo-constant if for any n-simplex $\alpha = (x_0 \overset{\alpha_1}{\leftarrow} \cdots \overset{\alpha_n}{\leftarrow} x_n)$ of Nerve($\mathbb{C}$), with corresponding map $\pi_\alpha : \Delta^n_\alpha \to B\mathbb{C}$, the sheaf A restricts to a pseudo-constant sheaf $\pi^*_\alpha(A)$ on Δ^n.

2.3. Lemma. *If A a pseudo-constant sheaf on Δ^n then $H^q(\Delta^n, A) = 0$ for each $q > 0$.*

Proof. By induction on n. The case $n = 0$ is clear. Suppose the lemma holds for $n - 1$. Let $\{U_i\}_{i=0}^\infty$ be a fundamental system of open neighbourhoods of the last face $\partial_n(\Delta^{n-1}) \subseteq \Delta^n$. Let $V = \Delta^n - \partial_n(\Delta^{n-1})$. For a suitable choice of the U_i, the sets V and $V \cap U_i$ are contractible, while the sheaf A is constant over V and $V \cap U_i$. So $H^q(V, A) = 0 = H^q(V \cap U_i, A)$ for $q > 0$. Thus the Mayer-Vietoris sequence for the cover $\Delta^n = V \cup U_i$ takes the form

$$\begin{aligned} 0 \to H^0(\Delta^n, A) \to\ & H^0(U_i, A) \oplus H^0(V, A) \to H^0(U_i \cap V, A) \to \\ \to H^1(\Delta^n, A) \to\ & H^1(U_i, A) \to 0 \\ .. \to H^q(\Delta^n, A) \to\ & H^q(U_i, A) \to 0 \ .. \ . \end{aligned}$$

Thus $H^q(\Delta^n, A) \to H^q(U_i, A)$ is an isomorphism, for each $q > 0$ and each i. But, by compactness of $\partial_n(\Delta^{n-1})$, one has $\varinjlim H^q(U_i, A) \cong H^q(\partial_n(\Delta^{n-1}), A)$, and the latter group vanishes for $q > 0$, by the induction hypothesis for Δ^{n-1}. Thus $H^q(\Delta^n, A) \cong \varinjlim H^q(U_i, A) = 0$ for $q > 0$, as required.

2.4. Corollary. *For any pseudo-constant sheaf A on $B\mathbf{C}$, there is a natural isomorphism $H^q(B\mathbf{C}, A) \tilde{\to} H^q(\Delta\mathbf{C}, \gamma A)$, for all $q \geq 0$.*

Analogous descriptions apply to topological categories. For a topological category $\mathbf{C}$, one can form the simplicial space Nerve($\mathbf{C}$), where a point of the space $\text{Nerve}_n(\mathbf{C})$ of n-simplices is denoted

$$\alpha = (x_0 \xleftarrow{\alpha_1} x_1 \leftarrow \cdots \xleftarrow{\alpha_n} x_n),$$

as before. The classifying space $B\mathbf{C}$ is defined by the (thickened) realization, as $B\mathbf{C} = |\text{Nerve}(\mathbf{C})|$.

For such a topological category $\mathbf{C}$, recall from Chapter II, Section 7 the topological category $\Delta_m(\mathbf{C})$ of simplices, and the associated continuous "first vertex" functor $\varphi : \Delta_m(\mathbf{C}) \to \mathbf{C}$.

Lemma 2.5. *For any topological category $\mathbf{C}$, the continuous functor $\varphi : \Delta_m(\mathbf{C}) \to \mathbf{C}$ induces a weak homotopy equivalence of the classifying spaces $B\Delta_m(\mathbf{C}) \tilde{\to} B\mathbf{C}$.*

The proof of this lemma is based on a suitable topological version of Quillen's Theorem A (Quillen(1973)). To state this version, let $\psi : \mathbf{D} \to \mathbf{C}$ be any continuous functor between topological categories. For an arbitrary map of spaces $f : X \to \mathbf{C}_0$, there is a new topological category ψ/X. Its objects are the triples (x, u, y) where $y \in \mathbf{D}_0, x \in X$ and $u : \psi(y) \to f(x)$ is an arrow in $\mathbf{C}$. Its arrows $\alpha : (x, u, y) \to (x', u', y')$ only exist if $x = x'$, and are arrows $\alpha : y \to y'$ in $\mathbf{D}$ for which $u' \circ \psi(\alpha) = u$. This category ψ/X is equipped with the evident topology: the space of objects $(\psi/X)_0$ is the fibered product $X \times_{\mathbf{C}_0} \mathbf{C}_1 \times_{\mathbf{C}_0} \mathbf{D}_0$, and the space of arrows $(\psi/X)_1$ can similarly be represented as a fibered product.

Now consider the special case where X is the space $\text{Nerve}_n(\mathbf{C})$ of n-simplices in a given topological category $\mathbf{C}$, and where $f : X \to \mathbf{C}_0$ is the "last vertex" map

$$\lambda_n : \text{Nerve}_n(\mathbf{C}) \to \mathbf{C}_0,$$

sending an n-simplex $(x_0 \leftarrow \cdots \leftarrow x_n)$ to x_n. Using the general properties of the thickened geometric realization mentioned in the previous section, Quillen's proof (op. cit.) now carries over verbatim, to show that if the given continuous functor $\psi : \mathbf{D} \to \mathbf{C}$ has the property that, for each $n \geq 0$, the evident projection

$$\varepsilon_n : B(\psi/\mathrm{Nerve}_n(\mathbf{C})) \to \mathrm{Nerve}_n(\mathbf{C})$$

is a weak homotopy equivalence, then so is $B\psi : B\mathbf{D} \to B\mathbf{C}$.

Proof of lemma. Apply these considerations to the particular functor $\varphi : \Delta_m(\mathbf{C}) \to \mathbf{C}$ in the statement of the lemma. For each $n \geq 0$, we may view $\mathrm{Nerve}_n(\mathbf{C})$ as a topological category with identity arrows only, and there are explicit continuous functors

$$\tau_n \circlearrowright \varphi/\mathrm{Nerve}_n(\mathbf{C}) \underset{\nu_n}{\overset{\varepsilon_n}{\rightleftarrows}} \mathrm{Nerve}_n(\mathbf{C}),$$

such that $\varepsilon_n \circ \nu_n = id$, and such that there are explicit continuous natural transformations $id \to \tau_n \leftarrow \nu_n \circ \varepsilon_n$, exactly as in the proof of Proposition II.7.6. By the homotopies produced by these natural transformations, the map $B(\varphi/\mathrm{Nerve}_n(\mathbf{C})) \to \mathrm{Nerve}_n(\mathbf{C})$, induced by ε_n, is a homotopy equivalence. Since this holds for each $n \geq 0$, the topological version of Quillen's Theorem A, just described, yields that $B\varphi : B\Delta_m(\mathbf{C}) \to B\mathbf{C}$ is a weak homotopy equivalence. This proves Lemma 2.5.

Recall from Chapter II, Section 3 the construction of the classifying topos $\mathcal{B}\mathbf{C}$ of $\mathbf{C}$-sheaves for a topological category $\mathbf{C}$, and its associated category $Ab(\mathcal{B}\mathbf{C})$ of abelian $\mathbf{C}$-sheaves. This applies in particular to the topological category $\Delta_m(\mathbf{C})$, to give an abelian category $Ab(\mathcal{B}\Delta_m(\mathbf{C}))$. For the category $Ab\ Sh(B\mathbf{C})$ of abelian sheaves on the classifying space, there is again a functor

$$\gamma : Ab\ Sh(B\mathbf{C}) \to Ab(\mathcal{B}\Delta_m(\mathbf{C}))\ , \tag{3}$$

similar to the one for a discrete category $\mathbf{C}$ in (2). Indeed, consider for each $n \geq 0$ the evident continuous maps

$$\begin{array}{ccc} \mathrm{Nerve}_n(\mathbf{C}) \times \Delta^n & \xrightarrow{\pi_n} & B\mathbf{C} \\ \downarrow{\scriptstyle p_n} & & \\ \mathrm{Nerve}_n(\mathbf{C}). & & \end{array} \tag{4}$$

Here p_n is simply the projection, while π_n is the map $X_n \times \Delta^n \to |X|^{(n)} \hookrightarrow |X|$ which exists for any simplicial space X. Then for each $n \geq 0$, one obtains a sheaf

$$A^{(n)} = (p_n)_* \, \pi_n^*(A)$$

on the space $\mathrm{Nerve}_n(\mathbf{C})$. These sheaves together define a sheaf

$$\gamma(A) =_{def} \sum_{n \geq 0} A^{(n)}$$

on the space $\sum_{n\geq 0}$ $\mathrm{Nerve}_n(\mathbb{C})$ of objects of $\Delta_m(\mathbb{C})$. Furthermore, this sheaf $\gamma(A)$ carries a natural continuous action (from the right) by the arrows of $\Delta_m(\mathbb{C})$. At the level of the stalks, this action is explicitly described as follows: Consider an arrow $u : (n,\alpha) \to (m,\beta)$ in $\Delta_m(\mathbb{C})$, where $\alpha = (x_0 \xleftarrow{\alpha_1} \cdots \xleftarrow{\alpha_n} x_n)$ and $\beta = (y_0 \xleftarrow{\beta_1} \cdots \xleftarrow{\beta_m} y_m)$; so $u : [n] \to [m]$ is a strictly monotone function with the property that $u^*(\beta) = \alpha$. Let $\Delta^n_\alpha \subseteq \mathrm{Nerve}_n(\mathbb{C}) \times \Delta^n$ be the copy of Δ^n corresponding to α, and similarly for $\Delta^m_\beta \subseteq \mathrm{Nerve}_m(\mathbb{C}) \times \Delta^m$. Then u induces an embedding $u : \Delta^n_\alpha \hookrightarrow \Delta^m_\beta$ for which $\pi_m \circ u = \pi_n$. Hence u yields a homomorphism

$$u^* : \Gamma(\Delta^m_\beta,\ \pi^*_m(A)) \to \Gamma(\Delta^n_\alpha,\ \pi^*_n(A)). \tag{5}$$

Since p_n in (4) is a proper map, the group $\Gamma(\Delta^m_\beta,\ \pi^*_m A)$ is precisely the stalk of $\gamma(A)$ at (m,β); and similarly for $\Gamma(\Delta^n_\alpha, \pi^*_n A)$. So (5) may alternatively be written as a homomorphism

$$u^* : \gamma(A)_{(m,\beta)} \to \gamma(A)_{(n,\alpha)}.$$

This defines the action by the arrow u on the sheaf $\gamma(A)$. It is readily verified that this action is continuous.

This construction, of the abelian $\Delta_m(\mathbb{C})$-sheaf $\gamma(A)$ from the abelian sheaf A on the space $B\mathbb{C}$, defines the functor γ announced in (3). It is clearly a left-exact functor. Its right-derived functors can be described in a way analogous to Lemma 2.1.

2.6. Lemma. *Let $\mathbb{C}$ be any topological category. For the stalks of the right-derived functors of the functor $\gamma : Ab\ Sh(B\mathbb{C}) \to Ab(\mathcal{B}\Delta_m(\mathbb{C}))$, there is for each abelian sheaf A on $B\mathbb{C}$ a natural isomorphism*

$$R^q\gamma(A)_{(n,\alpha)} \cong H^q(\Delta^n_\alpha, A),$$

for any point (n,α) in $\Delta_m(\mathbb{C})$. (On the right, A is identified with its restriction to $\Delta^n_\alpha \subseteq B\mathbb{C}$.)

Proof. Define for each $q \geq 0$ and each abelian sheaf A on $B\mathbb{C}$ an abelian $\Delta_m(\mathbb{C})$-sheaf $\mathcal{H}^q(A)$, as follows. For each $n \geq 0$, let $\mathcal{H}^q(A)^{(n)}$ be the sheaf on $\mathrm{Nerve}_n(\mathbb{C})$ defined as

$$\mathcal{H}^q(A)^{(n)} = R^q p_{n*}(\pi^*_n A), \tag{6}$$

where p_n and π_n are the maps described in (4). Since p_n is proper, the stalk of $\mathcal{H}^q(A)^{(n)}$ at a point (n,α) of $\mathrm{Nerve}_n(\mathbb{C})$ is given by

$$\mathcal{H}^q(A)^{(n)}_{(n,\alpha)} = H^q(\Delta^n_\alpha, A). \tag{7}$$

Much as for the construction of the $\Delta_m(\mathbb{C})$-sheaf A, one can now define an explicit action by $\Delta_m(\mathbb{C})$, on the sheaf $\mathcal{H}^q(A) := \sum_{n\geq 0} \mathcal{H}^q(A)^{(n)}$ over the space $\sum_{n\geq 0} \mathrm{Nerve}_n(\mathbb{C})$ of objects of $\Delta_m(\mathbb{C})$. This defines a functor $\mathcal{H}^q : Ab\ Sh(B\mathbb{C}) \to Ab(\mathcal{B}\Delta_m(\mathbb{C}))$, for each $q \geq 0$. Exactly as in Lemma 2.1, these functors $\mathcal{H}^q$ have the exactness and effacing properties which uniquely determine the right derived functors $R^q\gamma$. Therefore there is an isomorphism $\mathcal{H}^q(A) \cong R^q\gamma(A)$, natural in A. The isomorphism in the statement of the lemma now follows by (7).

Analogous to 2.4, one obtains the following immediate consequence.

2.7. Corollary. *For any topological category* $\mathbf{C}$ *and any pseudo-constant abelian sheaf* A *on the classifying space* $B\mathbf{C}$, *there is a natural isomorphism*

$$H^q(B\mathbf{C}, A) \cong H^q(\mathcal{B}\Delta_m(\mathbf{C}), \gamma(A)),$$

for each $q \geq 0$.

§3 Geometric realization by cosimplicial topoi

In this section we will consider analogous geometric realization functors, which take values in topoi rather than in spaces. This realization uses cosimplicial topoi. For example, if $Y^\cdot$ is a cosimplicial space (i.e., a covariant functor from $\mathbb{\Delta}$ into spaces), then $n \mapsto Sh(Y^n)$ is a cosimplicial topos.

If $\mathcal{D}^\cdot$ is a cosimplicial topos and X is a simplicial set, one can form a "tensor product"

$$|X|_{(\mathcal{D})} = X \otimes \mathcal{D}^\cdot$$

which can again be constructed by iterated pushouts, as

$$|X|_{(\mathcal{D})} = \varinjlim_k |X|^{(k)}_{(\mathcal{D})}, \tag{1}$$

where $|X|^{(k)}_{(\mathcal{D})}$ is defined from $|X|^{(k-1)}_{(\mathcal{D})}$ by the pushout of topoi (cf. Chapter I, Section 3)

$$\begin{array}{ccc} \sum_{x\in X_k^{(nd)}} \partial\mathcal{D}^k & \longrightarrow & \sum_{x\in X_k^{(nd)}} \mathcal{D}^k \\ \downarrow & & \downarrow \\ |X|^{(k-1)}_{(\mathcal{D})} & \longrightarrow & |X|^{(k)}_{(\mathcal{D})}. \end{array} \tag{2}$$

Here the boundary $\partial\mathcal{D}^k$ is constructed by forming suitable pushouts of faces in the category of topoi. For example, $\partial\mathcal{D}^1 = \mathcal{D}^0 + \mathcal{D}^0$, and $\partial\mathcal{D}^2$ is a pushout of three copies of $\mathcal{D}^1$, constructed by the two pushouts squares

$$\begin{array}{ccccc} \mathcal{D}^0 & \xrightarrow{\partial_1} & \mathcal{D}^1 & & \\ {\scriptstyle \partial^0}\downarrow & & \downarrow{\scriptstyle v} & & \\ \mathcal{D}^1 & \xrightarrow{u} & \mathcal{D}^1 \cup \mathcal{D}^1 & \longrightarrow & \partial\mathcal{D}^2 \\ & & \uparrow & & \uparrow \\ & & \mathcal{D}^0 + \mathcal{D}^0 & \longrightarrow & \mathcal{D}^1. \end{array}$$

This defines $|X|^{(k)}_{(\mathcal{D})}$ from $|X|^{(k-1)}_{(\mathcal{D})}$ for $k > 0$. The construction starts off by $|X|^{(0)}_{(\mathcal{D})} = \sum_{x \in X_0} \mathcal{D}^0$.

We will not use this construction for a general cosimplicial topos $\mathcal{D}^{\cdot}$, but only in the case where J is a topological interval as above with associated cosimplicial space $\Delta^{\cdot}_{(J)}$ of standard simplices, and $\mathcal{D}^{\cdot}$ is the associated cosimplicial topos $Sh(\Delta^{\cdot}_{(J)})$ of sheaves. For a simplicial set X, one thus obtains a *topos-theoretic realization with respect to the interval* J, denoted $\|X\|_J$:

$$\|X\|_J = X \otimes Sh(\Delta^{\cdot}_{(J)}). \tag{3}$$

(So the notations (1) and (3) are related by $\|X\|_J = |X|_{Sh(\Delta^{\cdot}_{(J)})}$.) This realization $\|X\|_J$ is thus explicitly constructed using colimits and pushouts of topoi, as

$$\|X\|_J = \varinjlim_k \|X\|^{(k)}_J \,, \tag{4}$$

where $\|X\|^{(0)}_J = \sum_{x \in X_0} Sh(\Delta^0_{(J)})$, and $\|X\|^{(k)}_J$ is constructed from $\|X\|^{(k-1)}_J$ as the pushout

$$\begin{array}{ccc} \sum_{x \in X^{(nd)}_k} \partial Sh(\Delta^k_{(J)}) & \longrightarrow & \sum_{x \in X^{(nd)}_k} Sh(\Delta^k_{(J)}) \\ \downarrow & & \downarrow \\ \|X\|^{(k-1)}_J & \longrightarrow & \|X\|^{(k)}_J . \end{array} \tag{5}$$

Note that $\|X\|_J$ depends functorially on J, just as the topological realization $|X|_{(J)}$ does (cf. Remark 1.1).

As an example, we consider this topos theoretic realization $\|X\|_\Sigma$ of a simplicial set X with respect to the Sierpinski interval Σ (cf. Example 1.2). Recall that

$$\Delta^0_{(\Sigma)} = pt, \quad \Delta^1_{(\Sigma)} = \Sigma \,, \quad \partial \Delta^2_{(\Sigma)} = \Delta^2_{(\Sigma)} \,, \dots .$$

Taking sheaves gives a cosimplicial topos

$$\mathcal{S} = Sh(\Delta^0_{(\Sigma)}) \,, \quad Sh(\Delta^1_{(\Sigma)}) \,, \quad Sh(\Delta^2_{\Sigma}), \dots$$

The realization using this cosimplicial topos doesn't collapse as quickly as its topological counterpart in Example 1.2, because the boundary operator ∂ does not commute with the operation of taking sheaves. Indeed, $Sh(\Delta^1_{(\Sigma)})$ is the category of triples $(E_0, E_1, \alpha : E_0 \to E_1)$, where E_0 and E_1 are sets and α is a function. Next, $\partial Sh(\Delta^2_{(\Sigma)})$ is the category whose objects are of the form $(E_0, E_1, E_2, \alpha_0, \alpha_1, \alpha_2)$ where α_i are functions as in

$$\begin{array}{ccccc} & & E_1 & & \\ & \overset{\alpha_2}{\nearrow} & & \overset{\alpha_0}{\searrow} & \\ E_0 & & \xrightarrow{\ \alpha_1\ } & & E_2, \end{array} \tag{6}$$

but this triangle need not commute. On the other hand, $Sh(\Delta^2_{(\Sigma)})$ is the category whose objects are *commuting* triangles of the form (6). For $n \geq 3$, one again has

$\partial Sh(\Delta^n_{(\Sigma)}) \cong Sh(\partial\Delta^n_{(\Sigma)}) = Sh(\Delta^n_{(\Sigma)})$. Note that this example also shows that $\partial Sh(\Delta^n_{(J)}) \to Sh(\Delta^n_{(J)})$ need not be an embedding of topoi.

Now let X be a simplicial set. In Example 1.2 we described the topological realization $|X|_{(\Sigma)}$ with respect to the Sierpinski interval. The topos theoretic realization is *not* the category of sheaves on this topological realization. To see this, let us compute $\|X\|_\Sigma$ by iterated pushouts. First, $\|X\|_\Sigma^{(0)}$ is the category of X_0-indexed sets, with typical object denoted $E = \{E_x\}_{x\in X_0}$. Next, $\|X\|_\Sigma^{(1)}$ fits into a pushout

$$\begin{array}{ccc}
\sum_{x\in X_1^{(nd)}} \partial Sh(\Delta^1_{(\Sigma)}) & \longrightarrow & \sum_{x\in X_1^{(nd)}} Sh(\Delta^1_{(\Sigma)}) \\
\downarrow & & \downarrow \\
\|X\|_\Sigma^{(0)} & \longrightarrow & \|X\|_\Sigma^{(1)}.
\end{array}$$

The topos on the upper right of this diagram is the category of $X_1^{(nd)}$-indexed families of sheaves on Σ, i.e. families of functions $F = \{\alpha_x : F_{x,0} \to F_{x,1}\}_{x\in X_1^{(nd)}}$. The topos on the upper left is the category $X_1^{(nd)}$-indexed families of pairs of sets $F = \{(F_{x,0}, F_{x,1})\}_{x\in X_1^{(nd)}}$. Thus a typical object of the topos $\|X\|_\Sigma^{(1)}$ is a triple (E, F, θ), where $E = \{E_x\}_{x\in X_0}$ is an indexed family of sets (an object of $\|X\|_\Sigma^{(0)}$), and $F = \{\alpha_x : F_{x,0} \to F_{x,1}\}_{x\in X_1^{(nd)}}$ is a family of arrows, while θ provides isomorphisms

$$\theta_{x,0} : F_{x,0} \cong E_{d_1 x}\ ,\ \theta_{x,1} : F_{x,1} \cong E_{d_0 x}\ ,$$

for each $x \in X_1^{(nd)}$. In other words, $\|X\|_\Sigma^{(1)}$ is equivalent to the category with as typical object a pair (E, α), where E is a family of sets $\{E_x\}_{x\in X_0}$ and α gives for each non-degenerate $x \in X_1$ a function

$$\alpha_x : E_{d_1 x} \to E_{d_0 x}\ .$$

(As a notational convention, we may, for $x \in X_1$ degenerate, define α_x to be the identity: $E_{d_1 x} = E_{d_0 x}$.) In the next stage, $\|X\|_\Sigma^{(2)}$ is constructed as the pushout

$$\begin{array}{ccc}
\sum \partial Sh(\Delta^2_{(\Sigma)}) & \longrightarrow & \sum Sh(\partial\Delta^2_{(\Sigma)}) \\
\downarrow & & \downarrow \\
\|X\|_\Sigma^{(1)} & \longrightarrow & \|X\|_\Sigma^{(2)}.
\end{array}$$

An explicit computation, based on the description of pushouts in Chapter I, Section 3, and similar to the computation of $\|X\|_\Sigma^{(1)}$ just given, shows that the pushout-topos $\|X\|_\Sigma^{(2)}$ is equivalent to the category of pairs (E, α), where $E = \{E_x\}_{x\in X_0}$ is an indexed family of sets, α is an indexed family of functions $\alpha_x : E_{d_1 x} \to E_{d_0 x}$ (all $x \in X_1$), where α_x = identity if x is degenerate, and moreover such that for any $y \in X_2^{(nd)}$ the triangle

$$\begin{array}{ccccc}
 & & E_{y_1} & & \\
 & {}^{\alpha_{d_2 y}}\nearrow & & \searrow^{\alpha_{d_0 y}} & \\
E_{y_0} & & \xrightarrow[\alpha_{d_1 y}]{} & & E_{y_2},
\end{array}$$

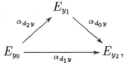

commutes. (Here y_0 denotes the zeroth vertex $y_0 = d_1d_1y = d_1d_2y$ of y, and similarly $y_1 = d_0d_2y = d_1d_2y$ and $y_2 = d_0d_0y = d_0d_1y$.) Since $\partial Sh(\Delta^k_{(\Sigma)}) = Sh(\Delta^k_{(\Sigma)})$ for $k \geq 3$, the sequence of iterated pushouts in the construction of $\|X\|_\Sigma$ stops here, and $\|X\|_\Sigma = \|X\|^{(2)}_\Sigma$ is the category of such triples (E, α).

For the special case where the simplicial set X is the nerve of a small category $\mathbb{C}$, with typical n-simplex of the form

$$c_0 \xleftarrow{f_1} c_1 \leftarrow \cdots \xleftarrow{f_n} c_n \ ,$$

the realization $\|\text{Nerve}(\mathbb{C})\|_\Sigma$ is precisely the category of contravariant functors from $\mathbb{C}$ into sets, as is clear from the explicit calculation of $\|X\|_\Sigma$ just described. We record this in the following theorem.

3.1. Theorem. *For any small category* $\mathbb{C}$*, there is a natural equivalence of topoi*

$$\|\text{Nerve}(\mathbb{C})\|_\Sigma \ \cong \ \mathcal{B}\mathbb{C}.$$

Just as for the ordinary geometric realization considered in the previous two sections, there is an analogous "thickened" topos theoretic realization $|X|_{\mathcal{D}^\cdot}$ for any simplicial space X and cosimplicial topos $\mathcal{D}^\cdot$, constructed as the tensor product of topoi $Sh(X_n) \otimes \mathcal{D}^n$. (It is thickened, in the sense that the tensor product is now taken over the subcategory $\mathbb{\Delta}_m \subseteq \mathbb{\Delta}$ consisting of injective functions only.) For the special case where $\mathcal{D}^\cdot = Sh(\Delta^\cdot_{(J)})$ for a topological interval J, we will again denote this topos by $\|X\|_J$. More concretely, and parallel to the case of simplicial sets, this topos $\|X\|_J$ is constructed as a colimit of topoi $\|X\|_J = \varinjlim \|X\|^{(k)}_J$, where $\|X\|^{(0)}_J = Sh(X_0)$, and where $\|X\|^{(k)}_J$ is constructed from $\|X\|^{(k-1)}_J$ as a pushout of topoi

$$\begin{array}{ccc} Sh(X_k) \times \partial Sh(\Delta^k_{(J)}) & \longrightarrow & Sh(X_k) \times Sh(\Delta^k_{(J)}) \\ \downarrow & & \downarrow \\ \|X\|^{(k-1)}_J & \longrightarrow & \|X\|^{(k)}_J , \end{array} \tag{7}$$

This is completely analogous to (3)-(5) for simplicial sets, except that, first, the topology of the spaces X_k is taken into account, and, secondly, we do not restrict to the subspaces $X^{(nd)}_k \subseteq X_k$ of non-degenerate k-simplices. These modifications are exactly the same as for the topological thickened geometric realization in Section 1.

As an example, consider again the case where J is the Sierpinski space Σ. For a simplicial space X, the calculation of the topos $\|X\|_\Sigma$ proceeds exactly as for the case of a simplicial set. In particular, for a topological category $\mathbb{C}$ one can apply this calculation to Nerve($\mathbb{C}$), to obtain the following result for the topos $\tilde{\mathcal{B}}\mathbb{C}$ of quasi-$\mathbb{C}$-sheaves, described in Chapter II, Section 7.

3.2. Theorem. *For any topological category* $\mathbb{C}$*, there is a natural equivalence of topoi*

$$\|\text{Nerve}(\mathbb{C})\|_\Sigma \ \cong \ \tilde{\mathcal{B}}\mathbb{C}\ .$$

Observe that, since $\|\mathrm{Nerve}(\mathbf{C})\|_\Sigma$ is defined as a "thickened" topos theoretic realization, the identity arrows in $\mathbf{C}$ are treated as ordinary arrows. This explains the occurrence in Theorem 3.2 of the "thickened" classifying topos $\bar{\mathcal{B}}\mathbf{C}$ instead of the standard one $\mathcal{B}\mathbf{C}$. Recall from Chapter II, Proposition 7.7 that $\mathcal{B}\mathbf{C}$ is a natural deformation retract of $\bar{\mathcal{B}}\mathbf{C}$.

§4 Sheaves and geometric realization

For a simplicial space X, one can construct the topos $Sh(|X|)$ of sheaves on the geometric realization of X, but one can also first take sheaves for each space X_n, and then take the topos theoretic realization (with respect to the standard unit interval I), as discussed in the previous section. The purpose of this section is to relate these two constructions. The topos theoretic realization $\|X\|_I$ will simply be denoted by $\|X\|$.

Let us consider the topos $\|X\|$ more closely. An object E of $\|X\|$ is a system $\{E_n : n \geq 0\}$ of sheaves, where each E_n is a sheaf on the product $X_n \times \Delta^n$, and these sheaves are required to be compatible, in the sense that for each n and each $i \in \{0, \cdots, n\}$, there is an isomorphism $(d_i \times 1)^*(E_{n-1}) \cong (1 \times \partial_i)^*(E_n)$; in other words, in the diagram below both squares are pullbacks.

$$\begin{array}{ccccc} E_{n-1} & \longleftarrow & \cdot & \hookrightarrow & E_n \\ \downarrow & & \downarrow & & \downarrow \\ X_{n-1} \times \Delta^{n-1} & \xleftarrow{d_i \times 1} & X_n \times \Delta^{n-1} & \xhookrightarrow{1 \times \partial_i} & X_n \times \Delta^n . \end{array} \tag{1}$$

Now let S be any sheaf on the realization $|X|$, i.e. an étale map $f : S \to |X|$. Consider for each $n \geq 0$ the canonical maps $u_n : X_n \times \Delta^n \twoheadrightarrow \|X\|^{(n)} \subseteq \|X\|$ (cf. (4) of Section 1). Then S pulls back to a sheaf $u_n^*(S)$ on $X_n \times \Delta^n$, and (by commutativity of Section 1, (4), and the definition of v_n there) these sheaves have the required compatibility property. In this way, we obtain a functor

$$Sh(|X|) \to \|X\|, \quad S \mapsto \{u_n^*(S)\}_{n \geq 0} \ .$$

This functor evidently commutes with colimits and finite limits, since each u_n does. Therefore, it is the inverse image, part of a topos morphism

$$\varphi : \|X\| \to Sh(|X|). \tag{2}$$

Thus, by definition,

$$\varphi^*(S)_n = u_n^*(S) \quad (n \leq 0). \tag{3}$$

Observe that the sheaf S can be reconstructed from these sheaves $u_n^*(S)$ on $X_n \times \Delta^n$. Explicitly, since $f : S \to |X|$ is an étale map and $|X|$ has the weak topology with respect to the filtration $|X|^{(0)} \subseteq |X|^{(1)} \subseteq |X|^{(2)} \subseteq \cdots$, it follows first that S has the weak topology with respect to its closed subspaces $S^{(0)} \subseteq S^{(1)} \subseteq S^{(2)} \subseteq \cdots$, where $S^{(n)} = f^{-1}(|X|^{(n)})$. Next, in the diagram

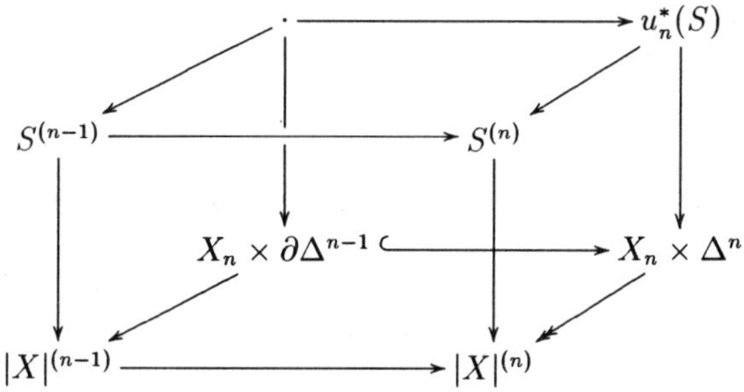

(4)

the bottom square is a pushout by construction of $|X|^{(n)}$, while all vertical faces are pullback squares. It follows that the top face (4) is also a pushout square. Indeed, since $X_n \times \Delta^n \to |X|^{(n)}$ is a quotient map, so is its pullback $u_n^*(S) \to S^{(n)}$ along the étale map $S^{(n)} \to |X|^{(n)}$. Thus it suffices to prove that the top face is a pushout of sets, which is easy.

It follows that for two sheaves S and T on $|X|$, a compatible family of sheaf maps $u_n^*(S) \to u_n^*(T)$ induces a unique continuous map $S \to T$ of sheaves on $|X|$. Thus, the functor φ^* in (3) is fully faithful; or, in other words, the topos morphism $\varphi : \|X\| \to Sh(|X|)$ is *connected.*

In the rest of this section, we will be concerned with the reverse construction, of a sheaf on $|X|$ from an object E of the topos $\|X\|$, i.e. from a compatible family of sheaves E_n on $X_n \times \Delta^n$. The construction is by pushouts similar to diagram (4). More precisely, from the sheaves E_n we construct a sequence of mappings $\tilde{E}^{(n)} \to |X|^{(n)}$, with the property that the two squares in the diagram

$$\begin{array}{ccccc} \tilde{E}^{n-1} & \longleftarrow & \cdot & \hookrightarrow & E_n \\ {\scriptstyle f_{(n-1)}}\downarrow & & \downarrow & & \downarrow \\ |X|^{(n-1)} & \longleftarrow & X_n \times \partial\Delta^n & \hookrightarrow & X_n \times \Delta^n . \end{array} \tag{5}$$

are pullbacks, as follows. For $n = 0$, define $\tilde{E}^{(0)} = E_0$ with evident map to $|X|^{(0)} = X_0$. Given $\tilde{E}^{(n-1)}$, define $\tilde{E}^{(n)}$ to be the pushout of the two top horizontal maps in (5). This defines a sequence of spaces and closed embeddings,

$$\tilde{E}^{(0)} \subseteq \tilde{E}^{(1)} \subseteq \tilde{E}^{(2)} \subseteq \cdots \quad . \tag{6}$$

Define $\tilde{E}$ to be the colimit of this sequence (i.e., the union, equipped with the weak topology). Then the $f_{(n)}$ together define a map $f : \tilde{E} \to |X|$.

The problem now is that this map $f : \tilde{E} \to |X|$ need not be étale, i.e. $\tilde{E}$ need not be a sheaf. We will prove that $\tilde{E}$ is a sheaf in two special cases: the first is when each X_n is a paracompact Hausdorff space, or briefly, when the simplicial space X is paracompact Hausdorff. The second special case is where the sheaves E_n on $X_n \times \Delta^n$ have a particularly simple form. The arguments for these two special cases are similar.

The first, paracompact Hausdorff, case is based on the following two lemmas. For the first lemma, fix a paracompact Hausdorff space Y, and an increasing sequence of closed subspaces

$$Y_0 \subseteq Y_1 \subseteq \cdots, \quad Y = \bigcup_{n\geq 0} Y_n .$$

Let $p : F \to Y$ be a continuous map, and write $F = \bigcup_n F_n$ where $F_n = p^{-1}(Y_n)$. Assume that Y and F carry the weak topology with respect to these filtrations $\{Y_n\}$ and $\{F_n\}$.

4.1. Lemma. *For a map $p : F \to Y$ and filtrations $Y = \bigcup Y_n$ and $F = \bigcup F_n$ as above, if each restriction $p_n = p|F_n : F_n \to Y_n$ is étale, then so is $p : F \to Y$.*

Proof. Clearly $p : F \to Y$ is an open map. For if $U \subseteq F$ is open, then $p(U) \cap Y_n = p(U \cap p^{-1}(Y_n)) = p_n(U \cap F_n)$ is open in Y_n because each $p_n : F_n \to Y_n$ is assumed open. Thus $p(U)$ is open in Y.

Next, to show that p has "enough sections", pick a point $\xi \in F$ and write $y = p(\xi)$. Fix the smallest n with $y \in Y_n$. By induction we will construct for each $k \geq n$ a neighbourhood U_k of y in Y_k and a section $s_k : \bar{U}_k \to F_k$ of p_k with $s_k(y) = \xi$ so that,

(i) $U_{k+1} \cap Y_k = U_k$,

(ii) $s_{k+1}|U_k = s_k : U_k \to F$.

Starting with $k = n$, let $r : V \to F_n$ be any section of p_n defined on an open neighbourhood V of y, with $r(y) = \xi$. (Such a section exists since p_n is assumed to be étale.) Let U_n be a neighbourhood of y with $y \in U_n \subseteq \bar{U}_n \subseteq V$, and let $s_n = r|\bar{U}_n$. By paracompactness of Y_{k+1}, the section $s_k : \bar{U}_k \to F_k \subseteq F_{k+1}$ can be extended to a section $\alpha : W \to F_{k+1}$ defined on an open neighbourhood W of $\bar{U}_k$ in Y_{k+1} (see Godement (1958), p. 150). Let W_1 be an open set in Y_{k+1} with $\bar{U}_k \subseteq W_1 \subseteq \bar{W}_1 \subseteq W$, and define $U_{k+1} = U_k \cup (W_1 - Y_k)$. This set is open in Y_{k+1}. Indeed, if $O \subseteq Y_{k+1}$ is any open set with $O \cap Y_k = U_k$, then $U_k = (O \cap W_1) \cap Y_k$, so $U_{k+1} = (O \cap W_1) \cup (W_1 - Y_k)$. Furthermore, since $\bar{U}_{k+1} \subseteq W$ we can define $s_{k+1} : \bar{U}_{k+1} \to F_{k+1}$ to be the restriction of α to $\bar{U}_{k+1}$. This completes the definition of the open sets $U_k \subseteq Y_k$ and the sections $s_k : \bar{U}_k \to F_k$ for all $k \geq n$. Now let

$$U = \bigcup_{k\geq n} U_k, \qquad s = \bigcup_{k\geq n} s_k : U \to F.$$

Then s is a continuous section of p, since $U \cap Y_k = U_k$ and $s|(U \cap Y_k) = s_k$ is continuous for each $k \geq n$. Furthermore, $s(U) \subseteq F$ is open, because for each $k \geq n$ the set $s(U) \cap F_k = s(U) \cap p^{-1}(Y_k) = s(U \cap s^{-1}p^{-1}(Y_k)) = s(U_k)$ is open in F_k. Then $p : s(U) \to U$ and $s : U \to s(U)$ are mutually inverse maps, so p must be a homeomorphism from the open neighbourhood $s(U)$ of ξ in F onto U. This shows that p is étale, and proves the lemma.

For the second lemma, consider, for a closed subspace $A \subseteq Y$ of a given space Y and a map $f : A \to B$, the adjunction space $Z = Y \cup_A B$. In other words, the square

$$\begin{array}{ccc} A & \xrightarrow{f} & B \\ \downarrow & & \downarrow \\ Y & \longrightarrow & Y \cup_A B = Z \end{array}$$

is a pushout. Then B is a closed subspace of Z, and the square is also a pullback (fibered product). A typical open set in Z is constructed by starting with an open $U \subseteq B$, and then choosing any open $V \subseteq Y$ with $V \cap A = f^{-1}(U)$. Then $V + U \subseteq Y + B$ is saturated for the equivalence relation $a \sim f(a)$ (for all $a \in A$) which defines Z as a quotient of $Y + B$. Hence the image $V \cup_A U$ of $V + U$ in Z is an open set. We will assume that Y and B are paracompact Hausdorff spaces. It then follows that Z is a paracompact Hausdorff space as well. (Hausdorffness of Z is easy; for paracompactness, see Michael (1957).)

4.2. Lemma. *For $A \subseteq Y$ and $f : A \to B$ as above and for any diagram*

$$\begin{array}{ccccc} F & \hookleftarrow & D & \xrightarrow{g} & G \\ {\scriptstyle r}\downarrow & & \downarrow{\scriptstyle p} & & \downarrow{\scriptstyle q} \\ Y & \hookleftarrow & A & \xrightarrow[f]{} & B \end{array}$$

in which both squares are pullbacks, if p, q, r are all étale maps, then so is the induced map

$$\pi = r \cup q \ : F \cup_D G \to Y \cup_A B$$

of adjunction spaces.

Proof. In the square

$$\begin{array}{ccc} F + G & \longrightarrow & F \cup_D G \\ {\scriptstyle r+q}\downarrow & & \downarrow{\scriptstyle \pi} \\ Y + B & \longrightarrow & X \cup_A B \end{array}$$

both horizontal maps are quotient maps; so π is a continuous open map since r and q are. To show that π is in fact a local homeomorphism, first note that B is a closed subspace of $Y \cup_A B$ and $(Y \cup_A B) - B = Y - A$. And similarly D is a closed subspace of $F \cup_D G$ and $F \cup_D G - E = F - D$. Thus π is a local homeomorphism over the open subset $(Y \cup_A B) - B$. It remains to be shown that each point $\xi \in \pi^{-1}(B)$ has an open neighbourhood V_ξ in Y such that $\pi | V_\xi$ is a homeomorphism $V_\xi \tilde{\to} \pi(V_\xi)$. To this end, choose such a point ξ, write $b = \pi(\xi) \in B$, and use étaleness of the map q to find an open neighbourhood U of b in B and a section $s : \bar{U} \to E$ through ξ. This section pulls back to a section $f^{\#}(s) : f^{-1}(\bar{U}) \to D$ of the map p. Since $f^{-1}(\bar{U})$ is closed in Y, we can now use paracompactness of Y to extend $f^{\#}(s)$ to a section $t : N \to E$ of

r on an open neighbourhood N of $f^{-1}(\bar{U})$. Now let $W = f^{-1}(U) \cup (N - A)$. Then W is an open subset of Y, and $W \cap A = f^{-1}(U)$. So, as noted before the statement of the lemma, W gives an open subset

$$V := W \cup_A U \subseteq Y \cup_A B.$$

Furthermore,

$$t \cup s : W \cup_A U \to F \cup_D G$$

is a well-defined section of π defined on V. Its image $t \cup s(V)$ is open in $F \cup_D G$, since $t \cup s(V) = t(W) \cup_D s(U)$ and $t(W), s(U)$ are open in F and G respectively, while $t(W) \cap D = t(W) \cap r^{-1}(A) = t(W \cap t^{-1}r^{-1}(A)) = t(W \cap A) = f^{\#}(s)(f^{-1}(U)) = g^{-1}(s(U))$. Thus $(t \cup s)(V)$ is the desired neighbourhood V_ξ of ξ on which π restricts to a homeomorphism, with inverse $t \cup s$.

4.3. Remark. Keeping the notation and the assumptions of the preceding lemma, both squares in the diagram

$$\begin{array}{ccccc} F & \longrightarrow & F \cup_D G & \longleftarrow & G \\ {\scriptstyle r}\downarrow & & \downarrow{\scriptstyle \pi} & & \downarrow{\scriptstyle q} \\ Y & \longrightarrow & Y \cup_A B & \longleftarrow & B \end{array}$$

are again pullback squares. Indeed, one readily verifies that these squares are set-theoretic pullbacks. But any commutative square of continuous maps

$$\begin{array}{ccc} S & \longrightarrow & T \\ \downarrow & & \downarrow \\ K & \longrightarrow & L \end{array}$$

which is a set-theoretic pullback and in which vertical maps are étale is also a topological pullback, since the map $S \to K \times_L T$ into topological pullback is a continuous bijection between étale spaces over K, hence a homeomorphism.

Using these two lemmas, one concludes that the construction of the map $\tilde{E} \to |X|$ from an object E of the topos $\|X\|$, described around (5) and (6) above, in fact results in an étale map, i.e. an object of $Sh(|X|)$. This shows that every object of $\|X\|$ is in the image of the functor $\varphi^* : Sh(|X|) \to \|X\|$ in (3) , and hence proves the following theorem. Recall that $\|X\|$ stands for the topos theoretic realization $\|X\|_I$ with respect to the standard unit interval I.

4.4. Theorem. *For any paracompact Hausdorff simplicial space X, the morphism $\varphi : \|X\| \to Sh(|X|)$ is an equivalence of topoi.*

This theorem applies in particular to any simplicial *set* X, viewed as a simplicial space with the discrete topology. For the nerve of a small (discrete) category $\mathbb{C}$, we state this explicitly as follows.

4.5. Corollary. *For any small (discrete) category* $\mathbb{C}$, *there is a canonical equivalence of topoi* $\|\mathrm{Nerve}(\mathbb{C})\| \xrightarrow{\sim} Sh(B\mathbb{C})$.

For application in the next chapter, we need to describe one more case where, for an object E of the topos $\|X\|$, the construction of $\tilde{E} \to |X|$ actually yields an étale map. Recall that such an E is a compatible system of sheaves E_n on $X_n \times \Delta^n$ (for $n \geq 0$). For each point $x \in X_n$ this sheaf thus restricts to a sheaf $E_n|(\{x\} \times \Delta^n)$ on the standard n-simplex. If, for each $n \geq 0$ and each point $x \in X_n$, this restricted sheaf is a *pseudo-constant* sheaf on Δ^n (cf. Section III.2) then we call E itself *pseudo-constant.* (Note, however, that these sheaves E_n on $X_n \times \Delta^n$ are allowed to vary arbitrarily in the X_n-coordinate.)

4.6. Proposition. *For any simplicial space* X, *each pseudo-constant object* E *of the topos* $\|X\|$ *is contained in the image of the functor* $\varphi^* : Sh(|X|) \to \|X\|$.

Proof. In the proof, we will, for any étale map $g : Z \to Y$, call an open set $U \subseteq Z$ *small* if g restricts to a homeomorphism on U. Exactly as in the proof of Theorem 4.4, we will show that the map $f : \tilde{E} \to |X|$ is a local homeomorphism; but now we use that E is pseudo-constant, rather than paracompactness of X. Recall that $\tilde{E}$ is constructed as a colimit of spaces $\tilde{E}^{(0)} \subseteq \tilde{E}^{(1)} \subseteq \cdots$ equipped with maps $\tilde{f}_{(n)} : \tilde{E}^{(n)} \to |X|^{(n)}$. Suppose, for the moment, that it has been shown that each of these maps $\tilde{E}^{(n)} \to |X|^{(n)}$ is étale; and that, moreover, for each each "small" neighbourhood U_{n-1} in $\tilde{E}^{(n-1)}$ there exists a small neighbourhood U_n in $\tilde{E}^{(n)}$ such that $U_n \cap \tilde{E}^{(n-1)} = \tilde{E}^{(n)}$. Then it will follow that $\tilde{E} \to |X|$ is étale, with small neighbourhoods of the form $\bigcup U_n$ for such a sequence $\{U_n\}$, exactly as in the proof of Lemma 4.1. It thus suffices to prove for each inclusion $\tilde{E}^{(n-1)} \subseteq \tilde{E}^{(n)}$ that if $\tilde{E}^{(n-1)} \to |X|^{(n-1)}$ is étale, then $\tilde{E}^{(n)} \to |X|^{(n)}$ is also étale and moreover has this extension property for small neighbourhoods.

For this, let $U = U_{n-1} \subseteq \tilde{E}^{(n-1)}$ be a small open set. Since both squares in (5) are pullbacks, U_{n-1} pulls back along $X_n \times \partial\Delta^n \to |X|^{(n-1)}$ to a small neighbourhood U' of $E_n|(X_n \times \partial\Delta^n)$. In other words, U' corresponds to a section s of E_n defined on an open subset V of $X_n \times \partial\Delta^n$. Let b be the barycenter of Δ^n, and define the (open) cone $C(V) \subseteq X_n \times \Delta^n$ to be the set of points $(x, t) \in X_n \times \Delta^n$ for which there are $\alpha \in [0, 1)$ and $t' \in \partial\Delta^n$ so that $(x, t') \in V'$ and $t = \alpha\, t' + (1 - \alpha)b$. Since E_n is assumed pseudo-constant, there is a unique extension of the section s to a section $\tilde{s}$ on $C(V)$. The image of $\tilde{s}$ defines an open set $W \subseteq E_n$. Consider now the pushout square defining $\tilde{E}^{(n)}$ from $\tilde{E}^{(n-1)}$ (cf. below (5)):

$$\begin{array}{ccc} E_n|(X_n \times \partial\Delta^n) & \xhookrightarrow{\;i\;} & E_n \\ {\scriptstyle a}\downarrow & & \downarrow{\scriptstyle b} \\ \tilde{E}^{(n-1)} & \xhookrightarrow{\;j\;} & \tilde{E}^{(n)} \end{array}$$

Then the open set W just defined has the property that $i^{-1}(W) = a^{-1}(U)$, and hence defines (by the description of the pushout-topology just before the statement of Lemma 4.2) a unique open set U_n of $\tilde{E}^{(n)}$ so that $j^{-1}(U_n) = U = U_{n-1}$ and $b^{-1}(U_n) = W$. Furthermore, exactly as in the proof of lemma 4.2, the map $\tilde{E}^{(n)} \to |X|^{(n)}$ restricts to a homeomorphism on U_n. Thus, as for Lemma 4.2, this proves that $\tilde{E}^{(n)} \to |X|^{(n)}$ is étale, and shows at the same time that the small open set $U_{n-1} \subseteq \tilde{E}^{(n-1)}$ can be extended to a small open set $U_n \subseteq \tilde{E}^{(n)}$, as required above.

This proves the proposition.

Chapter IV

Comparison Theorems

§1 Discrete categories

In this chapter, we will derive several theorems providing a homotopy theoretic comparison between classifying topoi and classifying spaces. We will begin with the relatively easy case of comparing the classifying space of a small category to the topos of presheaves on that category.

Let $\mathbb{C}$ be a small (discrete) category, with topos of presheaves $\mathcal{B}\mathbb{C}$ as described in Section 2 of Chapter I, and with classifying space $B\mathbb{C}$ as described in Section 2 of Chapter III. The general approach to geometric realization provides a map comparing these two constructions. Indeed, for the simplicial set Nerve($\mathbb{C}$) one can construct its topos-theoretic realization, both with respect to the standard unit interval $I = [0,1]$ and with respect to the Sierpinski interval Σ. For the first realization, Corollary 4.5 of the previous chapter states that

$$\|\mathrm{Nerve}(\mathbb{C})\|_I \cong Sh(B\mathbb{C}),$$

while for the second realization, Theorem 3.1 of that chapter states that

$$\|\mathrm{Nerve}(\mathbb{C})\|_\Sigma \cong \mathcal{B}\mathbb{C}.$$

The evident continuous map of intervals $p : I \to \Sigma$, defined by

$$p(t) = \begin{cases} 0 & , \quad t = 0, \\ 1 & , \quad t > 0, \end{cases}$$

thus induces a morphism of topoi, (again) denoted $p : \|\mathrm{Nerve}(\mathbb{C})\|_I \to \|\mathrm{Nerve}(\mathbb{C})\|_\Sigma$, or equivalently

$$p : B\mathbb{C} \to \mathcal{B}\mathbb{C}.$$

(Here we follow the convention in Section I.2 of identifying a space with its topos of sheaves.)

1.1. Theorem. *For any small category $\mathbb{C}$ this map $p : B\mathbb{C} \to \mathcal{B}\mathbb{C}$, from the classifying space to the classifying topos, is a weak homotopy equivalence of topoi.*

Before proving the theorem, we should give a more explicit description of the inverse image functor $p^* : \mathcal{B}\mathbb{C} \to Sh(B\mathbb{C})$ of the morphism p occurring in the statement of the theorem. For an object S of $\mathcal{B}\mathbb{C}$, i.e. a functor $S : \mathbb{C}^{op} \to (sets)$, one may picture the sheaf $p^*(S)$ on $B\mathbb{C}$ as built up in stages, following the filtration of $B\mathbb{C}$ by its skeleta $B\mathbb{C}^{(n)}$. The space $B\mathbb{C}^{(0)}$ is the set of objects of $\mathbb{C}$, equipped with the discrete topology. Then $p^*(S)^{(0)}$ is the sheaf on $B\mathbb{C}^{(0)}$ which has the set $S(c)$ as stalk over an object $c \in \mathbb{C}$. Next, for each non-identity arrow $\alpha : c_1 \to c_0$ there is a 1-simplex $\Delta^1_\alpha \subseteq B\mathbb{C}$, with endpoints c_0 and c_1. The restriction of $p^*(S)$ to this copy Δ^1_α is constant over $\Delta^1_\alpha - \{c_0\}$, with stalk $S(c_1)$, while the stalk $S(c_0)$ over c_0 is glued to this constant sheaf over $\Delta^1_\alpha - \{c_0\}$ via the map $S(\alpha) : S(c_0) \to S(c_1)$:

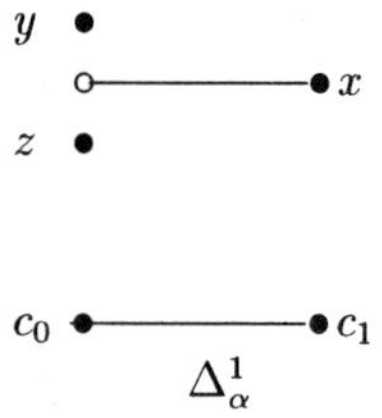

$x \in S(c_1); y, z \in S(c_0)$,
$S(\alpha)(y) = x = S(\alpha)(z)$.

Next, for a pair of (non-identity) arrows $\alpha = (c_0 \overset{\alpha_1}{\leftarrow} c_1 \overset{\alpha_2}{\leftarrow} c_2)$ there is a 2-simplex $\Delta^2_\alpha \subseteq B\mathbb{C}$, with vertices c_0, c_1, c_2 and faces corresponding to α_1, α_2 and $\alpha_1 \circ \alpha_2$. The restriction of the sheaf $p^*(S)$ to this 2-simplex Δ^2_α is constant over the complement in Δ^2_α of the face $\partial_2(\Delta^2_\alpha)$ (this is the face corresponding to α_1, opposite c_2), with stalk $S(c_2)$. Over the face $\partial_2(\Delta^2_\alpha) = \Delta^1_{\alpha_1}$ the sheaf $p^*(S)$ has already been described. These two parts are glued together to produce a sheaf on Δ^2_α, by using the restriction maps $S(c_0) \to S(c_2)$ and $S(c_1) \to S(c_2)$ given by $\alpha_1 \circ \alpha_2$ and by α_2. More generally, given the sheaf $p^*(S)^{(n-1)}$ on $B\mathbb{C}^{(n-1)}$, this sheaf is extended to a sheaf $p^*(S)^{(n)}$ on $B\mathbb{C}^{(n)}$ as follows: $B\mathbb{C}^{(n-1)}$ is a closed subspace of $B\mathbb{C}^{(n)}$, with inclusion map $i_n : B\mathbb{C}^{(n-1)} \hookrightarrow B\mathbb{C}^{(n)}$, say. Write Y_n for the open complement $B\mathbb{C}^{(n)} - B\mathbb{C}^{(n-1)}$, with inclusion map $j_n : Y_n \hookrightarrow B\mathbb{C}^{(n)}$. The space Y_n is a disjoint sum of interiors of n-simplices Δ^n_α, one for each non-degenerate n-simplex $\alpha = (c_0 \overset{\alpha_1}{\leftarrow} c_1 \leftarrow \cdots \overset{\alpha_n}{\leftarrow} c_n)$. Define a locally constant sheaf L_n on $Y_n = \bigcup_\alpha Int(\Delta^n_\alpha)$, which is constant over $Int(\Delta^n_\alpha)$ with stalk $S(c_n)$ (where c_n depends on α). Now glue this locally constant sheaf L_n on Y_n to the sheaf $p^*(S)^{(n-1)}$ already constructed, by "Artin glueing", using the map $p^*(S)^{(n-1)} \to i_{n*}j_n^*(L_n)$ defined in the evident way from the operators $S(\alpha_i \circ \cdots \circ \alpha_n)$: $S(c_i) \to S(c_n)$, for each $\alpha = (c_0 \overset{\alpha_1}{\leftarrow} \cdots \overset{\alpha_n}{\leftarrow} c_n)$ as above.

There are two properties of the sheaf $p^*(S)$ on $B\mathbb{C}$ that we will use. First, $p^*(S)$ is a pseudo-constant sheaf on $B\mathbb{C}$. Secondly, for any n-simplex $\alpha = (c_0 \leftarrow \cdots \leftarrow c_n)$ with associated map $\pi_\alpha : \Delta^n_\alpha \to B\mathbb{C}$ as in Section III.2, there is a natural isomorphism

$$\Gamma(\Delta^n_\alpha, \pi^*_\alpha(S)) \cong S(c_0). \tag{1}$$

These two properties are obvious from the description of $p^*(S)$ just given.

Proof of Theorem 1.1. If the category $\mathbb{C}$ splits into connected components as $\mathbb{C} = \sum \mathbb{C}_i$, then $B\mathbb{C}$ is the sum of the connected spaces $B\mathbb{C}_i$, while $\mathcal{B}\mathbb{C}$ is the sum of the connected topoi $\mathcal{B}\mathbb{C}_i$. From this it is clear that p induces an isomorphism $\pi_0(B\mathbb{C}) \cong \pi_0(\mathcal{B}\mathbb{C})$.

To prove that p induces an isomorphism of fundamental groups, it suffices to show that the functor $p^* : \mathcal{B}\mathbb{C} \to Sh(B\mathbb{C})$ restricts to an equivalence of categories on the full subcategories of $\mathcal{B}\mathbb{C}$ and $Sh(B\mathbb{C})$ consisting of locally constant objects. An object S of $\mathcal{B}\mathbb{C}$ is locally constant precisely when for each arrow $\alpha : c \to d$ in $\mathbb{C}$ the operator $S(\alpha) : S(d) \to S(c)$ is an isomorphism, i.e. S is morphism-inverting. And a sheaf E on $B\mathbb{C}$ is locally constant precisely when E is a covering projection. But there is a standard equivalence of categories $p_! : \{\textit{covering spaces of } B\mathbb{C}\} \to \{\textit{morphism-inverting functors } \mathbb{C}^{op} \to (\textit{sets})\}$, considered in Gabriel-Zisman(1967) and Quillen(1973). For a covering space $E \to B\mathbb{C}$, the functor $p_!(E) : \mathbb{C}^{op} \to (\textit{sets})$ sends an object c to the fiber of E over c (viewed as a 0-simplex of $B\mathbb{C}$); the action of an arrow $\alpha : c \to d$ in $\mathbb{C}$ on $p_!(E)$ is defined using path-lifting in E. One readily verifies that p^* and $p_!$ are mutually inverse functors, up to natural isomorphism, thus providing the required equivalence between categories of locally constant objects.

Next, we note that p induces isomorphisms in cohomology with locally constant coefficients. Let $A : \mathbb{C}^{op} \to \underline{Ab}$ be a morphism-inverting functor into the category of abelian groups. We claim that p induces an isomorphism $H^*(\mathcal{B}\mathbb{C}, A) \to H^*(B\mathbb{C}, p^*A)$. (Recall from Proposition II.6.1 that $H^*(\mathcal{B}\mathbb{C}, A)$ is the same as the cohomology $H^*(\mathbb{C}, A)$ of the category $\mathbb{C}$.) Consider the diagram of functors

$$\begin{array}{ccc} Ab(\mathcal{B}\mathbb{C}) & \xrightarrow{p^*} & Ab\, Sh(B\mathbb{C}) \\ & \underset{\varphi^*}{\searrow} & \downarrow{\scriptstyle \gamma} \\ & & Ab(\mathcal{B}(\Delta\mathbb{C})). \end{array}$$

Here γ is the functor defined in (2) of Section III.2, and φ^* is induced by the "first vertex" functor $\varphi : \Delta\mathbb{C} \to \mathbb{C}$ described there. By (1) above the diagram commutes, up to natural isomorphism. But $p^*(A)$ is a pseudo-constant abelian sheaf on $B\mathbb{C}$, so γ induces an isomorphism $H^n(B\mathbb{C}, p^*A) \to H^n(\mathcal{B}(\Delta\mathbb{C}), \gamma p^*A)$ as in Corollary III.2.4. If A is moreover locally constant (morphism-inverting), then φ^* also induces an isomorphism $H^*(\mathcal{B}\mathbb{C}, A) \to H^*(\mathcal{B}(\Delta\mathbb{C}), \varphi^*A)$; indeed, the Leray spectral sequence of Chapter II, Remark 6.3, collapses since each comma category φ/c is contractible (cf. (1) of Section III.2). By the isomorphism $\varphi^*A \cong \gamma p^*A$, we conclude that p induces an isomorphism $H^*(\mathcal{B}\mathbb{C}, A) \xrightarrow{\sim} H^*(B\mathbb{C}, p^*A)$, as claimed.

The theorem now follows from the toposophic Whitehead theorem stated in Section I.4.

Recall from Chapter II that for a small category $\mathbb{C}$ and a space X, the collection of concordance classes of principal $\mathbb{C}$-bundles is denoted $k_{\mathbb{C}}(X)$.

1.2. Corollary. *For any small category* $\mathbb{C}$ *and any CW-complex* X, *there is* *a natural isomorphism*

$$k_{\mathbb{C}}(X) \cong [X, B\mathbb{C}].$$

Of course, for a group G (viewed as a one-object category) this result specializes to the classical result that BG classifies principal G-bundles. For a monoid with cancellation, one recovers Segal's theorem (cf. Chapter II, Example 2.1(b)).

Proof. The corollary is a direct consequence of the existence of the weak homotopy equivalence of Theorem 1.1. Indeed, since $p : B\mathbb{C} \to \mathcal{B}\mathbb{C}$ induces isomorphisms of homotopy groups $\pi_n(B\mathbb{C}, x) \to \pi_n^{et}(\mathcal{B}\mathbb{C}, px)$ for any point x in $B\mathbb{C}$, a standard argument (using induction on the cells of X) shows that for any CW-complex X the map p induces an isomorphism

$$[X, B\mathbb{C}] \xrightarrow{\sim} [X, \mathcal{B}\mathbb{C}].$$

The corollary now follows by composing this isomorphism with that of Chapter II, Corollary 2.4.

§2 s-Etale categories

Recall that a topological category $\mathbb{C}$ is said to be s-étale if its source map $s : \mathbb{C}_1 \to \mathbb{C}_0$ is étale, i.e. a local homeomorphism. In this section, we will extend the comparison between the classifying space and the classifying topos (Theorem 1.1) to such s-étale categories.

For an s-étale category $\mathbb{C}$, the construction of a map $B\mathbb{C} \to \mathcal{B}\mathbb{C}$, from the classifying space to the classifying topos, is somewhat more involved then the construction for a discrete category in the previous section. Exactly as there, the functoriality of geometric realization provides a natural topos morphism

$$p : \|\text{Nerve}(\mathbb{C})\|_I \to \|\text{Nerve}(\mathbb{C})\|_\Sigma, \tag{1}$$

relating the realizations for the unit interval and for the Sierpinski interval. Furthermore, by Theorem 3.2 of Chapter II, the Sierpinski realization is the topos of quasi-$\mathbb{C}$-sheaves,

$$\|\text{Nerve}(\mathbb{C}\|_\Sigma \cong \tilde{\mathcal{B}}\mathbb{C}, \tag{2}$$

of which the classifying topos $\mathcal{B}\mathbb{C}$ is a natural deformation retract by Proposition 7.7 in Chapter II,

$$\tilde{\mathcal{B}}\mathbb{C} \underset{\tau}{\overset{\psi}{\rightleftarrows}} \mathcal{B}\mathbb{C}. \tag{3}$$

Next, by Theorem 4.4 of Chapter II the natural map

$$\varphi : \|\text{Nerve}(\mathbb{C})\|_I \to Sh(B\mathbb{C}) \tag{4}$$

is an equivalence if $\mathbb{C}$ is paracompact Hausdorff. Thus, for such a paracompact Hausdorff s-étale category $\mathbb{C}$, the above maps compose to give a canonical map $\bar{p}$ making the diagram

$$\begin{array}{ccc} Sh(B\mathbb{C}) & \overset{\bar{p}}{\dashrightarrow} & \mathcal{B}\mathbb{C} \\ {\scriptstyle \varphi}\uparrow & & \uparrow{\scriptstyle \psi} \\ \|\mathrm{Nerve}(\mathbb{C})\|_I & \xrightarrow[p]{} & \bar{\mathcal{B}}\mathbb{C} \end{array} \tag{5}$$

commute up to natural isomorphism. But even without the assumption that $\mathbb{C}$ is paracompact Hausdorff, there is a topos morphism $\bar{p} : Sh(B\mathbb{C}) \to \mathcal{B}\mathbb{C}$, unique up to isomorphism, such that $\psi p \cong \bar{p}\varphi$ as in (5). Indeed, from the explicit construction of the morphism p, it is not hard to see that every object in the image of the functor p^* is pseudo-constant (cf. the discussion following Theorem 1.1). Thus, by Proposition 4.6 of Chapter III, there is for every object S of $\bar{\mathcal{B}}\mathbb{C}$ a sheaf E on $B\mathbb{C}$ – unique up to isomorphism – with the property that $p^*S \cong \varphi^*E$. Since the morphism φ is connected (i.e. φ^* is full and faithful), a choice of such a sheaf E for every object S will give a functor $q^* : \bar{\mathcal{B}}\mathbb{C} \to Sh(B\mathbb{C})$ such that $\varphi^*q^* \cong p^*$. Since φ^* is faithful and φ^*q^* commutes with colimits and finite limits, so does q^*. Thus q^* is the inverse image of a topos morphism $q : Sh(B\mathbb{C}) \to \bar{\mathcal{B}}\mathbb{C}$ with the property that $q\varphi \cong p$. Now the morphism $\bar{p}$, defined as $\bar{p} = \psi q$, completes the diagram (5) as required.

We can now state and prove the analogue of Theorem 1.1. We assume that the topological category $\mathbb{C}$ is locally connected, so that, by Lemma 7.1 of Chapter II, the classifying topos $\mathcal{B}\mathbb{C}$ is also locally connected, as required for the construction of the étale homotopy groups (cf. Section I.1).

2.1. Theorem. *For any locally connected s-étale topological category $\mathbb{C}$, the natural morphism*

$$\bar{p} : B\mathbb{C} \to \mathcal{B}\mathbb{C}$$

is a weak homotopy equivalence.

Proof. The proof follows roughly the same pattern as for the discrete case, cf. Theorem 1.1. Thus it will be shown that $\bar{p}$ induces isomorphisms in π_0, in cohomology with locally constant coefficients, and in the fundamental group. The theorem then follows by the toposophic Whitehead Theorem.

First, that $\bar{p}$ induces an isomorphism $\pi_0(B\mathbb{C}) = \pi_0 Sh(B\mathbb{C}) \xrightarrow{\sim} \pi_0(\mathcal{B}\mathbb{C})$ is clear. Next, to see that $\bar{p}$ induces isomorphism in cohomology, note first that any abelian $\mathbb{C}$-sheaf A induces a pseudo-constant sheaf $\bar{p}^*(A)$, as in the construction of the morphism $\bar{p}$ above. With the notation of the previous section, it follows by Lemma 2.3 of Chapter III that $H^q(\Delta^n_\alpha, \bar{p}^*A) = 0$ for each $q > 0$, and for each point $\alpha \in \mathrm{Nerve}_n(\mathbb{C})$. Thus, by Corollary 2.4 of Chapter III, there is an isomorphism

$$H^q(B\mathbb{C}, \bar{p}^*A) \cong H^q(\mathcal{B}(\Delta_m\mathbb{C}), \gamma p^*A)$$

for any $q \geq 0$. If furthermore A is locally constant, then Proposition 7.6 of Chapter II gives an isomorphism $H^q(\mathcal{B}\mathbb{C}, A) \cong H^q(\mathcal{B}(\Delta_m\mathbb{C}), \varphi^*A)$. Since $\varphi^* \cong \gamma \circ \bar{p}^*$ as in the proof of Theorem 1.1, it follows that for such a locally constant A the topos morphism $\bar{p}$ induces an isomorphism $H^q(\mathcal{B}\mathbb{C}, A) \cong H^q(B\mathbb{C}, \bar{p}^*A)$.

Finally, we show that $\bar{p}$ induces an isomorphism $\pi_1(Sh(B\mathbb{C})) \to \pi_1(\mathcal{B}\mathbb{C})$, for any chosen (but not explicitly written) base-point in $B\mathbb{C}$. For this, it suffices to show that $\bar{p}^* : \mathcal{B}\mathbb{C} \to Sh(B\mathbb{C})$ restricts to an equivalence of categories between locally constant objects in $\mathcal{B}\mathbb{C}$ and covering spaces of $B\mathbb{C}$. To this end we give an explicit description of the locally constant objects in $\mathcal{B}\mathbb{C}$, viz. as the invertible $\mathbb{C}$-sheaves S for which the sheaf projection $p : S \to \mathbb{C}_0$ is a covering projection. Indeed, in the proof of Lemma 7.2 of Chapter II it was observed that if S is locally constant then S must be "invertible", since S restricts to a covering space $\mu^*(S)$ on $\Sigma \times \mathbb{C}_1$. The same argument shows that $p : S \to \mathbb{C}_0$ is a covering projection, since p is the pullback of $\mu^*(S)_0$ along the map $u : \mathbb{C}_0 \to \mathbb{C}_1$ which associates to each $x \in \mathbb{C}_0$ its identity arrow $u(x)$. Conversely, suppose that S is an invertible $\mathbb{C}$-sheaf for which p is a covering projection, say with fiber the set F. Then there is an étale surjection $\sigma : U \twoheadrightarrow \mathbb{C}_0$ for which there exists an isomorphism $\theta : F \times U \xrightarrow{\sim} S \times_{\mathbb{C}_0} U$ over U. To show that S is locally constant as a $\mathbb{C}$-sheaf, we need to produce a similar $\mathbb{C}$-equivariant isomorphism. To this end, consider the sheaf $U^{\mathbb{C}}$ whose points are pairs (y, α), where $y \in U$ and $\alpha : x \to \sigma(y)$ is some arrow in $\mathbb{C}$. In other words, $U^{\mathbb{C}}$ is the fibered product $U \times_{\mathbb{C}_0} \mathbb{C}_1$. This space $U^{\mathbb{C}}$ is a $\mathbb{C}$-sheaf, when equipped with the sheaf projection $s\pi_2 : U^{\mathbb{C}} \to \mathbb{C}_0$ sending a point (y, α) to the source $s(\alpha)$, and with action by $\mathbb{C}$ given by composition. (Note that the map $s\pi_2$ is indeed étale since $s : \mathbb{C}_1 \to \mathbb{C}_0$ and $\sigma : U \to \mathbb{C}_0$ are.) Now since S is invertible, the action induces an isomorphism

$$w : S \times_{\mathbb{C}_0} \mathbb{C}_1 \to \mathbb{C}_1 \times_{\mathbb{C}_0} S, \quad w(s, \alpha) = (\alpha, s \cdot \alpha).$$

The inverse of w may suggestively be written as

$$(\alpha, s) \quad \mapsto \quad (s \cdot \alpha^{-1}, \alpha).$$

(although there is no such thing as an arrow α^{-1}). One can now define the desired isomorphism of $\mathbb{C}$-sheaves

$$\rho : \Delta(F) \times U^{\mathbb{C}} \to S \times U^{\mathbb{C}}$$

(where the product is that of $\mathbb{C}$-sheaves), where $\Delta(F)$ is the constant $\mathbb{C}$-sheaf $F \times \mathbb{C}_0 \to \mathbb{C}_0$ with trivial $\mathbb{C}$-action, as follows: a point in $\Delta(F) \times U^{\mathbb{C}}$ is a quadruple (e, x, y, α) where $e \in F$, $x \in \mathbb{C}_0$, $y \in U$ and $\alpha : x \to \sigma(y)$; define

$$\rho(e, x, y, \alpha) = (\pi_1\theta(e, y) \cdot \alpha, (y, \alpha)).$$

This map ρ is $\mathbb{C}$-equivariant, since for any arrow $\beta : x' \to x$ in $\mathbb{C}$ we have

$$\begin{aligned} \rho((e, x, y, \alpha) \cdot \beta) &= \rho(e, x', y, \alpha \circ \beta) \\ &= (\pi_1\theta(e, y) \cdot \alpha\beta, (y, \alpha\beta)) \\ &= (\pi_1\theta(e, y) \cdot \alpha, (y, \alpha)) \cdot \beta \\ &= p(e, x, y, \alpha) \cdot \beta. \end{aligned}$$

Furthermore, ρ is an isomorphism, with inverse defined, for $s \in S$, $y \in U$ and $\alpha : p(s) \to y$, by

$$\rho^{-1}(s,(y,\alpha)) = (\theta^{-1}(s\cdot\alpha^{-1},y),p(s),y,\alpha).$$

With this explicit description of locally constant $\mathbb{C}$-sheaves, the equivalence between the category of such and the category of covering spaces of $B\mathbb{C}$ is clear: In one direction, the functor $\bar{p}^{\,*} : \mathcal{B}\mathbb{C} \to Sh(B\mathbb{C})$ sends locally constant $\mathbb{C}$-sheaves to covering projections of $B\mathbb{C}$, since any inverse image functor preserves locally constant objects. In the converse direction, any such covering space $E \to B\mathbb{C}$ pulls back along $\mathbb{C}_0 \hookrightarrow B\mathbb{C}$ to a covering space of $\mathbb{C}_0$, equipped with an invertible $\mathbb{C}$-action via the map $[0,1]\times\mathbb{C}_1 \to B\mathbb{C}$. This defines a functor $\bar{p}_!$ from covering spaces of $B\mathbb{C}$ into invertible $\mathbb{C}$-sheaves, i.e. into locally constant objects of the topos $\mathcal{B}\mathbb{C}$. These two functors $\bar{p}_!$, and $\bar{p}^{\,*}$ together provide the required equivalence of categories.

This completes the proof of the theorem.

2.2. Remark. For the case where $\mathbb{C}$ is an étale topological groupoid, as in II.4.4, Theorem 2.1 can be proved more easily, by a direct comparison of hypercovers; see Moerdijk(1991).

Call an s-étale category $\mathbb{C}$ *locally contractible* if its space $\mathbb{C}_0$ of objects (or equivalently, its space $\mathbb{C}_1$ of arrows) has a basis of contractible sets. Then the classifying space $B\mathbb{C}$ also has such a basis, and hence the homotopy groups of the space $B\mathbb{C}$ coincide with the étale homotopy groups of the topos $Sh(B\mathbb{C})$ (cf. Section I.4). In this case, Theorem 2.1 above and Corollary 4.3 of Chapter II together imply the following result, by exactly the same proof as for Corollary 1.2 in the previous section.

2.3. Corollary. *For any locally contractible s-étale category $\mathbb{C}$ and any CW-complex X, there is a natural bijection*

$$[X, B\mathbb{C}] \;\cong\; k_{\mathbb{C}}(X).$$

Thus, $B\mathbb{C}$ "classifies" concordance classes of principal $\mathbb{C}$-bundles.

§3 Segal's theorem on Γ^q

As an illustration of the use of the Comparison Theorem 2.1, we will present in this section a proof of Segal's theorem; cf. Segal(1978). (This proof is also described in Moerdijk(1991).) To state this theorem, let M be the monoid of smooth embeddings of $\mathbb{R}^q$ into itself. Thus M is a discrete category with just one object. Also, let Γ^q be the étale groupoid (cf. Chapter II, 4.4) with $\mathbb{R}^q$ as space of objects, while the arrows $x \to y$ in Γ^q are all germs of diffeomorphism $\varphi : U \to V$ where U and V are neighbourhoods of x and y respectively, and $\varphi(x) = y$. This set of arrows is equipped with the sheaf topology, so that the source and target maps of the groupoid Γ^q are étale. This groupoid plays an important role in the theory of foliations, since it "classifies" (in some sense) the smooth foliations of codimension q (cf. Haefliger(1984)).

3.1. Theorem. (Segal) *The classifying spaces $B\Gamma^q$ and BM are weakly homotopy equivalent.*

To prove Segal's theorem, we use Theorem 2.1, and prove instead the equivalent statement that the classifying topoi $\mathcal{B}\Gamma^q$ and $\mathcal{B}M$ are weakly homotopy equivalent. This turns out to be remarkably easy and explicit. We need the following auxiliary categories. Let $\tilde{\mathcal{D}}^q$ be the discrete category with open disks in $\mathbf{R}^q$ as objects and smooth embeddings as arrows. The monoid M is a subcategory of $\mathcal{D}^q$, and the inclusion

$$i : M \hookrightarrow \tilde{\mathcal{D}}^q$$

is an equivalence of categories, since every open disk is diffeomorphic to $\mathbf{R}^q$. The category $\tilde{\mathcal{D}}^q$ is a subcategory of the category of topological spaces, and the inclusion

$$Y \ : \ \tilde{\mathcal{D}}^q \hookrightarrow \text{(spaces)}$$

is a diagram of spaces on $\tilde{\mathcal{D}}^q$, in the sense of Chapter II, Section 5. Write

$$\mathcal{D}^q = Y_{\tilde{\mathcal{D}}^q}$$

for the associated s-étale topological category (as in Proposition II.5.1, with $\tilde{\mathcal{D}}^q$ for K). Thus, the space of objects of $\mathcal{D}^q$ is the disjoint sum of all open disks $W \subseteq \mathbf{R}^q$, and we denote an object of $\mathcal{D}^q$ as a pair (W, x), where x is a point in the disk W. An arrow $\alpha : (W, x) \to (V, y)$ in $\mathcal{D}^q$ is a smooth embedding $\alpha : W \hookrightarrow V$ with $\alpha(x) = y$. There is an obvious projection functor $\pi : \mathcal{D}^q \to \tilde{\mathcal{D}}^q$ (as in Section II.5), as well as an obvious functor $r : \mathcal{D}^q \to \Gamma^q$, defined on objects by $r(W, x) = x$ and on arrows by taking germs.

All these functors induce morphisms between classifying topoi, as in the diagram

$$\begin{array}{ccccccc} \Gamma^q & \overset{r}{\leftarrow} & \mathcal{D}^q & \overset{\pi}{\rightarrow} & \tilde{\mathcal{D}}^q & \overset{i}{\hookleftarrow} & M \\ \\ \mathcal{B}\Gamma^q & \overset{r}{\leftarrow} & \mathcal{B}\mathcal{D}^q & \overset{\pi}{\rightarrow} & \mathcal{B}\tilde{\mathcal{D}}^q & \cong & \mathcal{B}M. \end{array}$$

Here $i : \mathcal{B}\mathcal{D}^q \ \cong \ \mathcal{B}M$ is an equivalence of topoi, since $i : M \to \tilde{\mathcal{D}}^q$ is an equivalence of categories (see Section I.2). Furthermore, since each open disk is contractible, $\pi : \mathcal{B}\mathcal{D}^q \to \mathcal{B}\tilde{\mathcal{D}}^q$ is a weak homotopy equivalence by Chapter II, Corollary 6.9. Thus, the following proposition completes the proof of Theorem 3.1.

3.2. Proposition. *The topos morphism $r \ : \ \mathcal{B}\mathcal{D}^q \to \mathcal{B}\Gamma^q$ is a natural deformation retraction.*

More explicitly, this proposition asserts that there is a topos morphism $j : \mathcal{B}\Gamma^q \to$

$\mathcal{BD}^q$ such that $r \circ j \cong id$ while $j \circ r$ is "Sierpinski homotopic" to the identity (i.e. there is a natural transformation between the inverse image functors, cf. Chapter I, Section 4).

Proof of 3.2. By Theorem 4.1 of Chapter II, topos morphisms $\mathcal{B}\Gamma^q \to \mathcal{BD}^q$ correspond to Γ^q-equivariant principal $\mathcal{D}^q$-bundles over $\mathbf{R}^q$ (cf. Remark II.4.6). These are étale maps $E \to \mathbf{R}^q$ with a left $\mathcal{D}^q$-action which is principal, and a right Γ^q-action which respects the left $\mathcal{D}^q$-action. Define such a bundle E, in terms of the target map t of the groupoid Γ^q, by

$$E = \{(W, \alpha) \mid W \text{ an open disk in } \mathbf{R}^q, \alpha \text{ an arrow in } \Gamma^q, t(\alpha) \in W\}.$$

This space E is topologized as the disjoint sum of the subspaces $t^{-1}(W) \subseteq \Gamma^q$. The étale projection

$$s : E \to \mathbf{R}^q, \quad (W, \alpha) \mapsto s(\alpha)$$

makes E into a sheaf on $\mathbf{R}^q$. This sheaf is Γ^q-equivariant, by the obvious right Γ^q-action given by composition,

$$(W, \alpha) \cdot \gamma = (W, \alpha\gamma).$$

The space E has the structure of a $\mathcal{D}^q$-bundle, by the map

$$\pi : E \to \mathcal{D}_0^q \ , \quad \pi(W, \alpha) = (W, t(\alpha)),$$

and the left action of $\mathbf{D}^q$ by composition: for an arrow $\beta : (W, y) \to (V, z)$ in $\mathcal{D}^q$ with $y = t(\alpha)$,

$$\beta \cdot (W, \alpha) = (V, \beta\alpha). \tag{1}$$

To see that this bundle E defines a map

$$j : \mathcal{B}\Gamma^q \to \mathcal{BD}^q$$

(by $j^*(S) = S \otimes_{\mathcal{D}^q} E$), it suffices to check that the $\mathcal{D}^q$-action is principal. This is trivial. For example, condition (ii) for principality means that for any point $y \in \mathbf{R}^q$ and any two points (W, α) and (V, β) in E with $s(\alpha) = y = s(\beta)$, there is a third point (U, γ) in E with $s(\gamma) = y$, and arrows $\delta : (u, t\gamma) \to (W, t\alpha)$ and $\varepsilon : (U, t\gamma) \to (V, t\beta)$ in $\mathcal{D}^q$, such that, for the action (1), $\delta \cdot (U, \gamma) = (W, \alpha)$ and $\varepsilon \cdot (U, \gamma) = (V, \beta)$. To see that this condition holds, choose an open disk U around y so that the germs α and β are represented by embeddings $\alpha : U \hookrightarrow W$ and $\beta : U \hookrightarrow V$, and let γ be the identity germ at y, and let $\delta = \alpha$, $\varepsilon = \beta$.

To see that j is a map as required for the proposition, represent the map $r : \mathcal{BD}^q \to \mathcal{B}\Gamma^q$ by a $\mathcal{D}^q$-equivariant principal Γ^q-bundle. This is the bundle R defined by

$$R = \{(W, \beta) \mid W \text{ an open disk, } \beta \text{ an arrow in } \Gamma^q, \text{ and } s(\beta) \in W\}.$$

R is a $\mathcal{D}^q$-sheaf, with sheaf projection

$$s : R \to \mathcal{D}_0^q \ , \quad (W, \beta) \mapsto (W, s(\beta))$$

and right action given by composition in the category $\mathcal{D}^q$. And R has a principal Γ^q-action, defined by the map

$$t : R \to \mathbf{R}^q \qquad t(W,\beta) = t(\beta),$$

and left Γ^q-action given by composition in the groupoid Γ^q. To see that this bundle indeed defines the map $r : \mathcal{B}\mathcal{D}^q \to \mathcal{B}\Gamma^q$, observe that for any Γ^q-sheaf S and any object (W,x) in $\mathcal{D}^q$,

$$\begin{aligned}(S \otimes_{\Gamma^q} R)_{(W,x)} &\cong S \otimes_{\Gamma^q} R_{(W,x)}\\ &= S \otimes_{\Gamma^q} s^{-1}(x)\\ &= S_x = S_{r(W,x)} = r^*(S)_{(W,x)}.\end{aligned}$$

Now the composition $r \circ j : \mathcal{B}\Gamma^q \to \mathcal{B}\Gamma^q$ corresponds to the tensor product $R \otimes_{\mathcal{D}^q} E$, and there is an obvious isomorphism $c : R \otimes_{\mathcal{D}^q} E \to \Gamma^q$ given by composition in the groupoid Γ^q,

$$c((W,\beta) \otimes (W,\alpha)) = \beta\alpha.$$

Thus $r \circ j$ is isomorphic to the identity on $\mathcal{B}\Gamma^q$. Furthermore, for the composition $j \circ r : \mathcal{B}\mathcal{D}^q \to \mathcal{B}\mathcal{D}^q$, there is a natural transformation $id \to (jr)^* = r^*j^*$, corresponding to the map of $\mathcal{D}^q$-equivariant principal $\mathcal{D}^q$-bundles

$$g : \mathcal{D}^q \to E \otimes_{\Gamma^q} R,$$

which sends an arrow $\alpha : (W,x) \to (V,y)$ in $\mathcal{D}^q$ to its germ, or more precisely, to $(W,\alpha) \otimes (V,\alpha)$.

This completes the proof of Proposition 3.2, and hence also that of Theorem 3.1.

§4 Comparison for topological categories

In this section we will prove a comparison theorem for arbitrary topological categories. Following Section 5 of Chapter II, if a topological category $\mathbb{C}$ is not s-étale, we replace its "small" classifying topos $\mathcal{B}\mathbb{C}$ by the bigger Deligne classifying topos $\mathcal{D}\mathbb{C}$, which is the topos of sheaves on the simplicial space Nerve($\mathbb{C}$). More generally, for any simplicial space Y the following result compares the geometric realization $|Y|$ with the topos $Sh(Y)$ of sheaves on Y, introduced in Section II.5.

4.1. Theorem. *For any simplicial space Y, its geometric realization $|Y|$ has the same weak homotopy type as the topos $Sh(Y)$.*

In the proof of Theorem 4.1, we will use the auxiliary topological category Simp(Y) of simplices of Y. Its objects are pairs $([n],y)$ where $n \geq 0$ and $y \in Y_n$; its arrows $\alpha : ([n],y) \to ([m],z)$ are arrows $\alpha : [n] \to [m]$ in the simplicial category $\mathbb{\Delta}$ such that

$\alpha^*(z) = y$. This category Simp(Y) is topologized in the obvious way, similar to the topology of the categories $Y_{\mathbf{K}}$ introduced in Chapter II, Section 5. Note in this context that Simp(Y) is the dual of the category $Y_{\Delta^{op}}$.

The following lemma is well-known (see Segal(1974), Waldhausen(1983)), but for the convenience of the reader we have included a proof of it.

4.2. Lemma. *For any simplicial space Y, the geometric realization $|Y|$ has the same weak homotopy type as the classifying space BSimp(Y).*

Proof. We will use the basic property of realization of simplicial spaces, stated in Chapter II, Section 1, viz. that the realization of a map which is a weak homotopy equivalence in each degree is again a weak homotopy equivalence. The classifying space BSimp(Y) is the realization of the simplicial space Nerve(Simp(Y)), whose p-simplices can be written in the form

$$([n_0] \overset{\alpha_1}{\leftarrow} \cdots \overset{\alpha_p}{\leftarrow} [n_p], y) \;,\quad y \in Y_{n_0}. \tag{1}$$

Let T be the bisimplicial space whose p, q-simplices are of the form

$$([n_0] \overset{\alpha_1}{\leftarrow} \cdots \overset{\alpha_1}{\leftarrow} [n_p] \overset{\beta}{\leftarrow} [q], y)\,,\quad y \in Y_{n_0}. \tag{2}$$

The simplicial operators of T act in the p-direction as those of Nerve(Simp(Y)), and in the q-direction as those of the representable simplicial set $\Delta[n_p]$. There are obvious mappings

$$Y_q \overset{\lambda}{\longleftarrow} T_{p,q} \overset{\rho}{\longrightarrow} \text{Nerve}_p(\text{Simp}(Y));$$

the map λ sends a p, q-simplex as in (2) to $(\alpha_1 \circ \cdots \circ \alpha_p \circ \beta)^*(y)$, and ρ is simply defined by deleting β. It now suffices, by the basic property mentioned above, to show that λ induces a weak homotopy equivalence $|T_{\cdot,q}| \to Y_q$ for each q, and that ρ induces one $|T_{p,\cdot}| \to \text{Nerve}_p(\text{Simp}(Y))$ for each p. For a fixed q, the simplicial space $T_{\cdot,q}$ can be viewed as the nerve of the topological "comma" category $Y_q/\text{Simp}(Y)$. This category is related to the space Y_q (viewed as a topological category with identity arrows only) by obvious functors and natural transformations

$$Y_q \rightleftarrows Y_q/\text{Simp}(Y), \quad \begin{matrix} z \mapsto & ([q], z) & \overset{id}{\leftarrow} ([q], z) \\ z \leftarrow\!\shortmid & (([n], y) & \overset{\alpha}{\leftarrow} ([q], z)). \end{matrix}$$

This gives an explicit homotopy equivalence $|T_{\cdot,q}| = B(Y_q/\text{Simp}(Y)) \simeq Y_q$. For a fixed p, the space $|T_{p,\cdot}|$ is the disjoint sum $\sum_{[n_0]\leftarrow\cdots\leftarrow[n_p]} Y_{n_0} \times \Delta^q$, and the map $|T_{p,\cdot}| \to \text{Nerve}_p(\text{Simp}(Y) = \sum_{[n_0]\leftarrow\cdots\leftarrow[n_p]} Y_{n_0}$ induced by p is the projection, which is clearly a homotopy equivalence. This proves the lemma.

Proof of Theorem 4.1. Recall from Chapter II that Y can be viewed as a covariant diagram on the category Δ^{op}, and that $Sh(Y) \cong \mathcal{B}(Y_{\Delta^{op}})$ as in Proposition II.5.1. The topological category $Y_{\Delta^{op}}$ is s-étale, so Theorem 2.1 provides a weak homotopy equivalence

$$B(Y_{\Delta^{op}}) \to Sh(Y).$$

But $B(Y_{\Delta^{op}})$ is homeomorphic to the classifying space of the dual category $(Y_{\Delta^{op}})^{op}$, which is exactly the category Simp(Y) of simplices of Y. By the preceding lemma, it thus follows that $B(Y_{\Delta^{op}})$ has the same weak homotopy type as the realization $|Y|$.

Let us call a simplicial space Y locally contractible if each Y_n has a basis of contractible sets. Recall that for a space X, the collection of concordance classes of linearly ordered sheaves on X with an augmentation into Y is denoted by $Lin_c(X, Y)$.

4.3 Corollary. *For any locally contractible simplicial space Y and any CW-complex X, there is a natural bijection*

$$[X, |Y|] \cong Lin_c(X, Y).$$

Proof. This follows from Theorem 4.1 and Corollary II.5.6, exactly as for Corollary 2.3.

For the special case where Y is the nerve of a topological category $\mathbb{C}$, we state these results explicitly as follows.

4.4. Corollary. *For any topological category $\mathbb{C}$, the classifying space $B\mathbb{C}$ has the same weak homotopy type as the Deligne topos $\mathcal{D}\mathbb{C}$.*

Thus we can transfer Corollary II.5.8 to topological spaces, to obtain the following result:

4.5. Corollary. *Let $\mathbb{C}$ be a locally contractible topological category. Then $B\mathbb{C}$ classifies (concordance classes of) $\mathbb{C}$-augmented linear orders, in the sense that there is a natural bijection*

$$[X, B\mathbb{C}] \cong Lin_c(X, \mathbb{C}),$$

for any CW-complex X.

References

M. Artin, B. Mazur, *Etale Homotopy*, Springer LNM **100**, Springer-Verlag, Berlin, 1969.

M. Artin, A. Grothendieck, and J.-L. Verdier, *Théorie de topos et cohomologie étale des schémas*, ("SGA4") Springer LNM **269** and **270**, Springer-Verlag, Berlin, 1972.

R. Bott, Lectures on characteristic classes and foliations, in Springer LNM **279** (1972), 1-94.

A.K. Bousfield, D.M. Kan, *Homotopy Limits, Completions, and Localizations*, Springer LNM 304 (1972).

P. Deligne, Théorie de Hodge, III, *Publ. Math. IHES*, **44** (1975), 5-77.

R. Diaconescu, Change-of-base for toposes with generators, *J. Pure Appl. Alg.* **6** (1975), 191-218.

C.H. Dowker, D. Strauss, Sums in the category of frames, *Houston J. Math.* **3** (1977), 7-15.

R. Fritsch, R.A. Piccinini, *Cellular Structures in Topology*, Cambridge University Press, 1990.

P. Gabriel, M. Zisman, *Calculus of Fractions and Homotopy Theory*, Ergebnisse der Math. **35**, Springer-Verlag, Berlin, 1967.

R. Godement, *Topologie Algébrique et Théorie des Faisceaux*, Hermann, Paris, 1958.

A. Grothendieck, Sur quelques points d'algébre homologique, *Tohoku Math. J.* **9** (1957), 119-221.

A. Grothendieck, *Revêtements Etales et Groupe Fondamental* ("SGA1"), Springer LNM **224**, Springer-Verlag, Berlin, 1971.

A. Haefliger, Structures feuilletées et cohomologie à valeur dans un faisceau de groupoïdes, *Comm. Math. Helv.* **32** (1958), 248-329.

A. Haefliger, Groupoïde d'holonomie et classifiants, *Astérisque* **116** (1984), 183-194.

L. Illusie, *Complexe Cotangent et Déformations II*, Springer LNM **283**, Springer-Verlag, Berlin, 1972.

J.R. Isbell, Product spaces in locales, *Proc. Amer. Math. Soc.* **81** (1981), 116-118.

B. Iversen, *Sheaf Cohomology*, Springer-Verlag, New York, **1986**.

A. Joyal, M. Tierney, *An Extension of the Galois Theory of Grothendieck*, Mem. A.M.S. **309** (1984).

S. Mac Lane, *Categories for the Working Mathematician*, Springer-Verlag, New York, 1971.

S. Mac Lane, I. Moerdijk, *Sheaves in Geometry and Logic – A first introduction to topos theory*, Springer-Verlag (1992).

M. Makkai, R. Paré, *Accessible Categories: The Foundations of Categorical Model Theory*, Contemp. Math. **104** (1989).

J.P. May, *Simplicial Objects in Algebraic Topology*, Van Nostrand, New York, 1967 (reprinted by University of Chicago Press, Chicago, 1982).

E. Michael, Another note on paracompact spaces, *Proc. Amer. Math. Soc.* **8** (1957), 822-828.

J.S. Milne, *Etale Cohomology*, Princeton University Press, Princeton, 1980.

J.W. Milnor, The geometric realization of a semi-simplicial complex, *Ann. Math.* **65** (1957), 357-362.

I. Moerdijk, Continuous fibrations and inverse limits of toposes, *Compositio Math.* **58** (1986), 45-72.

I. Moerdijk, The classifying topos of a continuous groupoid, I, *Trans. A.M.S.* **310** (1988), 629-668.

I. Moerdijk, Prodiscrete groups and Galois toposes, *Indag. Math. (= Proc. Kon. Ned. Ac. Wet. series A)* **51** (1989), 219-235.

I. Moerdijk, Classifying toposes and foliations, *Ann. Inst. Fourier* **41** (1991), 189-209.

D. Quillen, *Homotopical Algebra*, Springer LNM 43, Springer-Verlag, Berlin, 1967.

D. Quillen, Higher K-theory I, in *Algebraic K-theory I*, (ed. H. Bass), Springer LNM **341**, Springer-Verlag, Berlin, 1973, 85-147.

G.B. Segal, Classifying spaces and spectral sequences, *Publ. Math. I.H.E.S.* **34** (1968), 105-112.

G.B. Segal, Categories and cohomology theories, *Topology* **13** (1974), 239-312.

G.B. Segal, Classifying spaces related to foliations, *Topology* **17** (1978), 367-382.

R.G. Swan, *The Theory of Sheaves*, University of Chicago Press, Chicago, 1964.

F. Waldhausen, Algebraic K-theory of spaces, in Springer LNM 1126 (1983), 318-419.

Index

Vol. 1526: J. Azéma, P. A. Meyer, M. Yor (Eds.), Séminaire de Probabilités XXVI. X, 633 pages. 1992.

Vol. 1527: M. I. Freidlin, J.-F. Le Gall, Ecole d'Eté de Probabilités de Saint-Flour XX – 1990. Editor: P. L. Hennequin. VIII, 244 pages. 1992.

Vol. 1528: G. Isac, Complementarity Problems. VI, 297 pages. 1992.

Vol. 1529: J. van Neerven, The Adjoint of a Semigroup of Linear Operators. X, 195 pages. 1992.

Vol. 1530: J. G. Heywood, K. Masuda, R. Rautmann, S. A. Solonnikov (Eds.), The Navier-Stokes Equations II – Theory and Numerical Methods. IX, 322 pages. 1992.

Vol. 1531: M. Stoer, Design of Survivable Networks. IV, 206 pages. 1992.

Vol. 1532: J. F. Colombeau, Multiplication of Distributions. X, 184 pages. 1992.

Vol. 1533: P. Jipsen, H. Rose, Varieties of Lattices. X, 162 pages. 1992.

Vol. 1534: C. Greither, Cyclic Galois Extensions of Commutative Rings. X, 145 pages. 1992.

Vol. 1535: A. B. Evans, Orthomorphism Graphs of Groups. VIII, 114 pages. 1992.

Vol. 1536: M. K. Kwong, A. Zettl, Norm Inequalities for Derivatives and Differences. VII, 150 pages. 1992.

Vol. 1537: P. Fitzpatrick, M. Martelli, J. Mawhin, R. Nussbaum, Topological Methods for Ordinary Differential Equations. Montecatini Terme, 1991. Editors: M. Furi, P. Zecca. VII, 218 pages. 1993.

Vol. 1538: P.-A. Meyer, Quantum Probability for Probabilists. X, 287 pages. 1993.

Vol. 1539: M. Coornaert, A. Papadopoulos, Symbolic Dynamics and Hyperbolic Groups. VIII, 138 pages. 1993.

Vol. 1540: H. Komatsu (Ed.), Functional Analysis and Related Topics, 1991. Proceedings. XXI, 413 pages. 1993.

Vol. 1541: D. A. Dawson, B. Maisonneuve, J. Spencer, Ecole d' Eté de Probabilités de Saint-Flour XXI - 1991. Editor: P. L. Hennequin. VIII, 356 pages. 1993.

Vol. 1542: J.Fröhlich, Th.Kerler, Quantum Groups, Quantum Categories and Quantum Field Theory. VII, 431 pages. 1993.

Vol. 1543: A. L. Dontchev, T. Zolezzi, Well-Posed Optimization Problems. XII, 421 pages. 1993.

Vol. 1544: M.Schürmann, White Noise on Bialgebras. VII, 146 pages. 1993.

Vol. 1545: J. Morgan, K. O'Grady, Differential Topology of Complex Surfaces. VIII, 224 pages. 1993.

Vol. 1546: V. V. Kalashnikov, V. M. Zolotarev (Eds.), Stability Problems for Stochastic Models. Proceedings, 1991. VIII, 229 pages. 1993.

Vol. 1547: P. Harmand, D. Werner, W. Werner, M-ideals in Banach Spaces and Banach Algebras. VIII, 387 pages. 1993.

Vol. 1548: T. Urabe, Dynkin Graphs and Quadrilateral Singularities. VI, 233 pages. 1993.

Vol. 1549: G. Vainikko, Multidimensional Weakly Singular Integral Equations. XI, 159 pages. 1993.

Vol. 1550: A. A. Gonchar, E. B. Saff (Eds.), Methods of Approximation Theory in Complex Analysis and Mathematical Physics IV, 222 pages, 1993.

Vol. 1551: L. Arkeryd, P. L. Lions, P.A. Markowich, S.R. S. Varadhan. Nonequilibrium Problems in Many-Particle Systems. Montecatini, 1992. Editors: C. Cercignani, M. Pulvirenti. VII, 158 pages 1993.

Vol. 1552: J. Hilgert, K.-H. Neeb, Lie Semigroups and their Applications. XII, 315 pages. 1993.

Vol. 1553: J.-L- Colliot-Thélène, J. Kato, P. Vojta. Arithmetic Algebraic Geometry. Trento, 1991. Editor: E. Ballico. VII, 223 pages. 1993.

Vol. 1554: A. K. Lenstra, H. W. Lenstra, Jr. (Eds.), The Development of the Number Field Sieve. VIII, 131 pages. 1993.

Vol. 1555: O. Liess, Conical Refraction and Higher Microlocalization. X, 389 pages. 1993.

Vol. 1556: S. B. Kuksin, Nearly Integrable Infinite-Dimensional Hamiltonian Systems. XXVII, 101 pages. 1993.

Vol. 1557: J. Azéma, P. A. Meyer, M. Yor (Eds.), Séminaire de Probabilités XXVII. VI, 327 pages. 1993.

Vol. 1558: T. J. Bridges, J. E. Furter, Singularity Theory and Equivariant Symplectic Maps. VI, 226 pages. 1993.

Vol. 1559: V. G. Sprindžuk, Classical Diophantine Equations. XII, 228 pages. 1993.

Vol. 1560: T. Bartsch, Topological Methods for Variational Problems with Symmetries. X, 152 pages. 1993.

Vol. 1561: I. S. Molchanov, Limit Theorems for Unions of Random Closed Sets. X, 157 pages. 1993.

Vol. 1562: G. Harder, Eisensteinkohomologie und die Konstruktion gemischter Motive. XX, 184 pages. 1993.

Vol. 1563: E. Fabes, M. Fukushima, L. Gross, C. Kenig, M. Röckner, D. W. Stroock, Dirichlet Forms. Varenna, 1992. Editors: G. Dell'Antonio, U. Mosco. VII, 245 pages. 1993.

Vol. 1564: J. Jorgenson, S. Lang, Basic Analysis of Regularized Series and Products. IX, 122 pages. 1993.

Vol. 1565: L. Boutet de Monvel, C. De Concini, C. Procesi, P. Schapira, M. Vergne. D-modules, Representation Theory, and Quantum Groups. Venezia, 1992. Editors: G. Zampieri, A. D'Agnolo. VII, 217 pages. 1993.

Vol. 1566: B. Edixhoven, J.-H. Evertse (Eds.), Diophantine Approximation and Abelian Varieties. XIII, 127 pages. 1993.

Vol. 1567: R. L. Dobrushin, S. Kusuoka, Statistical Mechanics and Fractals. VII, 98 pages. 1993.

Vol. 1568: F. Weisz, Martingale Hardy Spaces and their Application in Fourier Analysis. VIII, 217 pages. 1994.

Vol. 1569: V. Totik, Weighted Approximation with Varying Weight. VI, 117 pages. 1994.

Vol. 1570: R. deLaubenfels, Existence Families, Functional Calculi and Evolution Equations. XV, 234 pages. 1994.

Vol. 1571: S. Yu. Pilyugin, The Space of Dynamical Systems with the C^0-Topology. X, 188 pages. 1994.

Vol. 1572: L. Göttsche, Hilbert Schemes of Zero-Dimensional Subschemes of Smooth Varieties. IX, 196 pages. 1994.

Vol. 1573: V. P. Havin, N. K. Nikolski (Eds.), Linear and Complex Analysis – Problem Book 3 – Part I. XXII, 489 pages. 1994.

Vol. 1574: V. P. Havin, N. K. Nikolski (Eds.), Linear and Complex Analysis – Problem Book 3 – Part II. XXII, 507 pages. 1994.

Vol. 1575: M. Mitrea, Clifford Wavelets, Singular Integrals, and Hardy Spaces. XI, 116 pages. 1994.

Vol. 1576: K. Kitahara, Spaces of Approximating Functions with Haar-Like Conditions. X, 110 pages. 1994.

Vol. 1577: N. Obata, White Noise Calculus and Fock Space. X, 183 pages. 1994.

Vol. 1578: J. Bernstein, V. Lunts, Equivariant Sheaves and Functors. V, 139 pages. 1994.

Vol. 1579: N. Kazamaki, Continuous Exponential Martingales and *BMO*. VII, 91 pages. 1994.

Vol. 1580: M. Milman, Extrapolation and Optimal Decompositions with Applications to Analysis. XI, 161 pages. 1994.

Vol. 1581: D. Bakry, R. D. Gill, S. A. Molchanov, Lectures on Probability Theory. Editor: P. Bernard. VIII, 420 pages. 1994.

Vol. 1582: W. Balser, From Divergent Power Series to Analytic Functions. X, 108 pages. 1994.

Vol. 1583: J. Azéma, P. A. Meyer, M. Yor (Eds.), Séminaire de Probabilités XXVIII. VI, 334 pages. 1994.

Vol. 1584: M. Brokate, N. Kenmochi, I. Müller, J. F. Rodriguez, C. Verdi, Phase Transitions and Hysteresis. Montecatini Terme, 1993. Editor: A. Visintin. VII. 291 pages. 1994.

Vol. 1585: G. Frey (Ed.), On Artin's Conjecture for Odd 2-dimensional Representations. VIII, 148 pages. 1994.

Vol. 1586: R. Nillsen, Difference Spaces and Invariant Linear Forms. XII, 186 pages. 1994.

Vol. 1587: N. Xi, Representations of Affine Hecke Algebras. VIII, 137 pages. 1994.

Vol. 1588: C. Scheiderer, Real and Étale Cohomology. XXIV, 273 pages. 1994.

Vol. 1589: J. Bellissard, M. Degli Esposti, G. Forni, S. Graffi, S. Isola, J. N. Mather, Transition to Chaos in Classical and Quantum Mechanics. Montecatini Terme, 1991. Editor: S. Graffi. VII, 192 pages. 1994.

Vol. 1590: P. M. Soardi, Potential Theory on Infinite Networks. VIII, 187 pages. 1994.

Vol. 1591: M. Abate, G. Patrizio, Finsler Metrics – A Global Approach. IX, 180 pages. 1994.

Vol. 1592: K. W. Breitung, Asymptotic Approximations for Probability Integrals. IX, 146 pages. 1994.

Vol. 1593: J. Jorgenson & S. Lang, D. Goldfeld, Explicit Formulas for Regularized Products and Series. VIII, 154 pages. 1994.

Vol. 1594: M. Green, J. Murre, C. Voisin, Algebraic Cycles and Hodge Theory. Torino, 1993. Editors: A. Albano, F. Bardelli. VII, 275 pages. 1994.

Vol. 1595: R.D.M. Accola, Topics in the Theory of Riemann Surfaces. IX, 105 pages. 1994.

Vol. 1596: L. Heindorf, L. B. Shapiro, Nearly Projective Boolean Algebras. X, 202 pages. 1994.

Vol. 1597: B. Herzog, Kodaira-Spencer Maps in Local Algebra. XVII, 176 pages. 1994.

Vol. 1598: J. Berndt, F. Tricerri, L. Vanhecke, Generalized Heisenberg Groups and Damek-Ricci Harmonic Spaces. VIII, 125 pages. 1995.

Vol. 1599: K. Johannson, Topology and Combinatorics of 3-Manifolds. XVIII, 446 pages. 1995.

Vol. 1600: W. Narkiewicz, Polynomial Mappings. VII, 130 pages. 1995.

Vol. 1601: A. Pott, Finite Geometry and Character Theory. VII, 181 pages. 1995.

Vol. 1602: J. Winkelmann, The Classification of Three-dimensional Homogeneous Complex Manifolds. XI, 230 pages. 1995.

Vol. 1603: V. Ene, Real Functions – Current Topics. XIII, 310 pages. 1995.

Vol. 1604: A. Huber, Mixed Motives and their Realization in Derived Categories. XV, 207 pages. 1995.

Vol. 1605: L. B. Wahlbin, Superconvergence in Galerkin Finite Element Methods. XI, 166 pages. 1995.

Vol. 1606: P.-D. Liu, M. Qian, Smooth Ergodic Theory of Random Dynamical Systems. XI, 221 pages. 1995.

Vol. 1607: G. Schwarz, Hodge Decomposition – A Method for Solving Boundary Value Problems. VII, 155 pages. 1995.

Vol. 1608: P. Biane, R. Durrett, Lectures on Probability Theory. VII, 210 pages. 1995.

Vol. 1609: L. Arnold, C. Jones, K. Mischaikow, G. Raugel, Dynamical Systems. Montecatini Terme, 1994. Editor: R. Johnson. VIII, 329 pages. 1995.

Vol. 1610: A. S. Üstünel, An Introduction to Analysis on Wiener Space. X, 95 pages. 1995.

Vol. 1611: N. Knarr, Translation Planes. VI, 112 pages. 1995.

Vol. 1612: W. Kühnel, Tight Polyhedral Submanifolds and Tight Triangulations. VII, 122 pages. 1995.

Vol. 1613: J. Azéma, M. Emery, P. A. Meyer, M. Yor (Eds.), Séminaire de Probabilités XXIX. VI, 326 pages. 1995.

Vol. 1614: A. Koshelev, Regularity Problem for Quasilinear Elliptic and Parabolic Systems. XXI, 255 pages. 1995.

Vol. 1615: D. B. Massey, Lê Cycles and Hypersurface Singularities. XI, 131 pages. 1995.

Vol. 1538 – Second Edition: P.-A. Meyer, Quantum Probability for Probabilists. X, 312 pages. 1995.